Penguin Special
A Blueprint for Survival

A Blueprint for Survival

By the editors of *The Ecologist*

Penguin Books

Penguin Books Ltd, Harmondsworth,
Middlesex, England
Penguin Books Australia Ltd, Ringwood,
Victoria, Australia

First published as Vol. 2, No. 1 of *The Ecologist*, 1972
Revised edition published in Penguin Specials, 1972
Reprinted 1973

Copyright © *The Ecologist*, 1972

Made and printed in Great Britain by
Richard Clay (The Chaucer Press) Ltd,
Bungay, Suffolk
Set in Monotype Plantin

This book is sold subject to the condition that
it shall not, by way of trade or otherwise, be lent,
re-sold, hired out, or otherwise circulated without
the publisher's prior consent in any form of
binding or cover other than that in which it is
published and without a similar condition
including this condition being imposed on the
subsequent purchaser

Contents

Preface 9
Statement of Support 11
Acknowledgements 14

1 Introduction: The Need for Change 15
2 Towards the Stable Society: Strategy for Change 30
3 The Goal 62
Appendix A: Ecosystems and their disruption 69
Appendix B: Social systems and their disruption 94
Appendix C: Populations and food supply 117
Appendix D: Non-renewable resources 129

The Movement for Survival 134
References 136

Preface

This document has been drawn up by a small team of people, all of whom, in different capacities, are professionally involved in the study of global environmental problems.

Four considerations have prompted us to do this:

1. An examination of the relevant information available has impressed upon us the extreme gravity of the global situation today. For, if current trends are allowed to persist, the breakdown of society and the irreversible disruption of the life-support systems on this planet, possibly by the end of the century, certainly within the lifetimes of our children, are inevitable.

2. Governments, and ours is no exception, are either refusing to face the relevant facts, or are briefing their scientists in such a way that their seriousness is played down. Whatever the reasons, no corrective measures of any consequence are being undertaken.

3. This situation has already prompted the formation of the Club of Rome, a group of scientists and industrialists from many countries which is currently trying to persuade governments, industrial leaders and trade unions throughout the world to face these facts and to take appropriate action while there is yet time. It must now give rise to a national movement to act at a national level, and if need be to assume political status and contest the next general election. It is hoped that such an example will be emulated in other countries, thereby giving rise to an international movement, complementing the invaluable work being done by the Club of Rome.

4. Such a movement cannot hope to succeed unless it has previously formulated a new philosophy of life, whose goals

can be achieved without destroying the environment, and a precise and comprehensive programme for bringing about the sort of society in which it can be implemented.

This we have tried to do, and our *Blueprint for Survival* heralds the formation of the MOVEMENT FOR SURVIVAL (see p. 135) and, it is hoped, the dawn of a new age in which Man will learn to live with the rest of Nature rather than against it.

The Ecologist
Edward Goldsmith, Robert Allen, Michael Allaby, John Davoll, Sam Lawrence.

Statement of Support

The undersigned, without endorsing every detail, fully support the basic principles embodied in this *Blueprint for Survival*, both in respect of the analysis of the problems we face today and in the solutions proposed.

Professor Don Arthur, M.Sc., Ph.D., D.Sc., F.I.Biol., *Professor of Zoology, King's College, London*

Professor D. Bryce-Smith, D.Sc., *Professor of Organic Chemistry, University of Reading*

Sir Macfarlane Bunet, O.M., F.R.S., M.D., Sc.D., F.R.C.P., *School of Microbiology, University of Melbourne*

Professor C. F. Cornford, Hon.A.R.C.A., *Royal College of Art, London*

Sir Frank Fraser Darling

Professor G. W. Dimbleby, B.Sc., M.A., D.Phil., *Professor of Human Environment, Institute of Archaeology, London*

Professor George Dunnet, B.Sc., Ph.D., *Professor of Zoology, University of Aberdeen*

Dr P. N. Edmunds, B.Sc., M.D., M.R.C.Path., *Department of Bacteriology, Fife District Laboratory*

Professor R. W. Edwards, D.Sc., F.I.Biol., *Professor of Applied Biology, University of Wales Institute of Science and Technology*

Dr S. R. Eyre, B.Sc., Ph.D., *Department of Geography, University of Leeds*

Professor Douglas Falconer, B.Sc., Ph.D., F.I.Biol., *Professor of Genetics, University of Edinburgh*

Professor John Friend, B.Sc., Ph.D., F.I.Biol., *Professor of Botany, University of Hull*

Professor F. W. Grimes, C.B.E., D.Litt., F.F.A., F.M.A., *Institute of Archaeology, University of London*

Professor John Hawthorn, B.Sc., Ph.D., F.R.S.E., F.R.I.C., F.I.F.S.T., *Professor of Food Science, University of Strathclyde*

Professor G. Melvyn Howe, M.Sc., Ph.D., *Professor of Geography, University of Strathclyde*

Sir Julian Huxley, F.R.S.

Dr David Lack, D.Sc., F.R.S., *Reader in Ornithology, Edward Grey Institute of Field Ornithology, University of Oxford*

Dr J. P. Lester, *British Medical Association*

Dr John A. Loraine, D.Sc., M.B., Ph.D., F.R.C.P.End., *Director, MRC Clinical Endocrinology Unit, Edinburgh*

Diana G. M. Loraine, R.G.M., S.C.M.

Professor L. F. Lowenstein, M.A., Dip.Psych., Ph.D., *Senior Educational Psychologist, School of Psychological Service, Hants.*

Dr Aubrey Manning, B.Sc., D.Phil., *Reader in Zoology, University of Edinburgh*

Professor Vincent Marks, *Professor of Biology, University of Surrey*

Sir Peter Medawar, C.H., F.R.S., *Nobel Prize Laureate, Medical Research Council (former Director of M.R.C.)*

Professor Ivor Mills, Ph.D., M.D., F.R.C.P., *Professor of Medicine, Department of Investigative Medicine, University of Cambridge*

Dr E. Mishan, Ph.D., *Reader in Economics, London School of Economics, and Professor of Economics, American University, Washington*

Professor P. J. Newbold, B.A., Ph.D., F.I.Biol., *Professor of Biology, The New University of Ulster*

Professor the Marquess of Queensberry, Hon. Des.R.C.A., M.S.I.A., *Royal College of Art*

Professor Forbes W. Robertson, Ph.D., D.Sc., F.I.Biol., *Professor of Genetics, University of Aberdeen*

Professor W. A. Robson, B.Sc. Econ., L.L.M., Ph.D., D.Litt., D. de L'Université, *Professor Emeritus in Public Administration, London School of Economics*

Dr J. Rose, M.Sc., Ph.D., F.I.L., F.R.I.C., *Director, Institute of Environmental Sciences, and Editor,* International Journal of Environmental Sciences

Sir Edward Salisbury, F.R.S.

Dr R. Scorer, M.A., Ph.D., F.R.S.H., F.I.M.A., *Imperial College, London,* and member of the Clean Air Council

Peter Scott, C.B.E., D.Sc., Ll.D., *Hon. Director of the Wild Fowl Trust*

Dr Malcolm Slesser, B.Sc., Ph.D., *Department of Pure and Applied Chemistry, University of Strathclyde*

Professor C. H. Waddington, C.B.E., F.R.S., *Professor of Animal Genetics, University of Edinburgh*

Dr A. Watson, D.Sc., Ph.D., F.R.S.E.

Professor V. C. Wynne-Edwards, F.R.S., *Regius Professor of Natural History, University of Aberdeen, and former Chairman, Natural Environment Research Council*

Acknowledgements

We would like to acknowledge the valuable comments contributed by Gerald Leach, The Rt Rev. Hugh Montefiore, Brian Johnson and John Papworth.

We are grateful to Potomac Associates, Washington D.C., for permission to reproduce four graphs from their forthcoming book *The Limits of Growth*, by Dennis Meadows; to the MIT Press for permission to use a number of tables and to quote extensively from their book *Man's Impact on the Global Environment*, *The Study of Critical Environmental Problems* (*SCEP*); to Pemberton Books for permission to reproduce a graph from their book *Population and Liberty*, by Jack Parsons; to Collier-Macmillan for permission to reproduce two tables from their book *Too Many*, by Georg Borgstrom; to Tom Stacey for permission to quote extensively from his book *Can Britain Survive?*, edited by E. Goldsmith.

Parts of the Introduction and 'Towards the Stable Society', notably those sections on stabilizing the population and on creating a new social system, have been adapted from a book by Robert Allen (to be published in 1973 by Allen Lane The Penguin Press) by permission of author and publisher.

1 Introduction: the Need for Change

The principal defect of the industrial way of life with its ethos of expansion is that it is not sustainable. Its termination within the lifetime of someone born today is inevitable – unless it continues to be sustained for a while longer by an entrenched minority at the cost of imposing great suffering on the rest of mankind. We can be certain, however, that sooner or later it will end (only the precise time and circumstances are in doubt) and that it will do so in one of two ways: either against our will, in a succession of famines, epidemics, social crises and wars; or because we want it to – because we wish to create a society which will not impose hardship and cruelty upon our children – in a succession of thoughtful, humane and measured changes. We believe that a growing number of people are aware of this choice, and are more interested in our proposals for creating a sustainable society than in yet another recitation of the reasons why this should be done. We will therefore consider these reasons only briefly, reserving a fuller analysis for the four appendices which follow the *Blueprint* proper.

Radical change is both necessary and inevitable because the present increases in human numbers and *per capita* consumption, by disrupting ecosystems and depleting resources, are undermining the very foundations of survival. At present the world population of 3,600 million is increasing by 2 per cent per year (72 million), but this overall figure conceals crucially important differences between countries. The industrialized countries with one third of the world population have annual growth rates of between 0·5 and 1·0 per cent; the undeveloped countries on the other hand, with two thirds of the world population, have annual growth rates of between 2 and 3 per cent, and from 40 to 45 per cent of their populations is under 15. It is commonly overlooked that in countries with an unbalanced age structure of this kind the population will continue to increase for many years even after fertility has fallen to the replacement level. As the Population Council has pointed out: 'If replacement is achieved in the developed world by 2000 and in the developing world by 2040, then the world's population will stabilise at nearly 15·5 billion (15,500 million) about a century hence, or well over four times the present size.'

The *per capita* use of energy and raw materials also shows a sharp division between the developed and the undeveloped parts of the world. Both are increasing their use of these commodities, but consumption in the developed countries is so much higher that, even with their smaller share of the population, their consumption may well represent over 80 per cent of the world total. For the same reason, similar percentage increases are far more significant in the developed countries; to take one example, between 1957 and 1967 *per capita* steel consumption rose by 12 per cent in the US and by 41 per cent in India, but the actual increases (in kg per year) were from 568 to 634 and from 9·2 to 13 respectively. Nor is there any sign that an eventual end to economic growth is envisaged, and indeed industrial economies appear to break down if growth ceases or even slows, however high the absolute level of consumption. Even the US still aims at an annual growth of GNP of 4 per cent or more. Within this overall figure much higher growth rates occur for the use of particular resources, such as oil.

The combination of human numbers and *per capita* consumption has a considerable impact on the environment, in terms of both the resources we take from it and the pollutants we impose on it. A distinguished group of scientists, who came together for a Study of Critical Environmental Problems (SCEP) under the auspices of the Massachusetts Institute of Technology, state in their report the clear need for a means of measuring this impact, and have coined the term 'ecological demand', which they define as 'a summation of all man's demands on the environment, such as the extraction of resources and the return of wastes'. Gross Domestic Product (GDP), which is population multiplied by material standard of living, appears to provide the

most convenient measure of ecological demand, and according
to the UN *Statistical Yearbook* this is increasing annually by 5
to 6 per cent, or doubling every 13·5 years. If this trend should
continue, then in the time taken for world population to double
(which is estimated to be by just after the year 2000) total
ecological demand will have increased by a factor of six. SCEP
estimate that 'such demand-producing activities as agriculture,
and mining and industry have global annual rates of increase of 3·5
per cent and 7 per cent respectively. An integrated rate of increase
is estimated to be between 5 and 6 per cent per year, in comparison with an annual rate of population increase of only 2 per
cent.'

It should go without saying that the world cannot accommodate this continued increase in ecological demand. *Indefinite*
growth of whatever type cannot be sustained by *finite* resources.
This is the nub of the environmental predicament. It is still less
possible to maintain indefinite *exponential* growth – and unfortunately the growth of ecological demand is proceeding
exponentially (i.e. it is increasing geometrically, by compound
interest).

The implications of exponential growth are not generally
appreciated and are well worth considering. As Professor
Forrester explains it (1):*

... pure exponential growth possesses the characteristic of behaving
according to a 'doubling time'. Each fixed time interval shows a
doubling of the relevant system variable. Exponential growth is
treacherous and misleading. A system variable can continue through
many doubling intervals without seeming to reach significant size.
But then in one or two more doubling periods, still following the same
law of exponential growth, it suddenly seems to become overwhelm-
ing.

Thus, supposing world petroleum reserves stood at 2,100
billion barrels, and supposing our rate of consumption was
increasing by 6·9 per cent per year, then, as can be seen from
Figure 1, demand will exceed supply by the end of the century.
What is significant, however, is not the speed at which such vast

reserves can be depleted, but that as late as 1975 there will appear to be reserves fully ample enough to last for considerably longer. Such a situation can easily lull one into a false sense of security and the belief that a given growth rate can be sustained, if not indefinitely, at least for a good deal longer than is actually the case.† The same basic logic applies to the availability of any

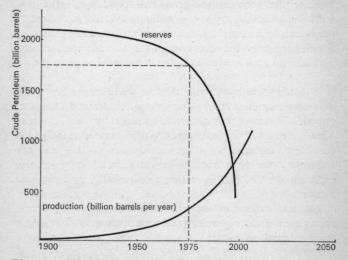

Figure 1   World reserves of crude petroleum at exponential rate of consumption. Note that in 1975, with no more than fifteen years left before demand exceeds supply, the total global reserve has been depleted by only $12\frac{1}{2}$ per cent.

resource including land, and it is largely because of this particular dynamic of exponential growth that the environmental predicament has come upon us so suddenly, and why its solution requires urgent and radical measures, many of which run

† It is perhaps worth bearing in mind that the actual rate of petroleum consumption *is* increasing by 6·9 per cent per year, and according to the optimistic estimate of W. P. Ryman, Deputy Exploration Manager of the Standard Oil Company of New Jersey, world petroleum reserves (including deposits yet to be discovered) are about 2,100 billion barrels.

counter to values which, in our industrial society, we have been taught to regard as fundamental.

If we allow the present growth rate to persist, total ecological demand will increase by a factor of 32 over the next 66 years – and there can be no serious person today willing to concede the possibility, or indeed the desirability, of our accommodating the pressures arising from such growth. For this can be done only at the cost of disrupting ecosystems and exhausting resources, which must lead to the failure of food supplies and the collapse of society. It is worth briefly considering each in turn.

DISRUPTION OF ECOSYSTEMS

We depend for our survival on the predictability of ecological processes. If they were at all arbitrary, we would not know when to reap or sow, and we would be at the mercy of environmental whim. We could learn nothing about the rest of nature, advance no hypotheses, suggest no 'laws'. Fortunately, ecological processes *are* predictable, and, although theirs is a relatively young discipline, ecologists have been able to formulate a number of important 'laws', one of which in particular relates to environmental predictability: namely, that all ecosystems tend towards stability, and further that the more diverse and complex the ecosystem the more stable it is; that is, the more species there are, and the more they interrelate, the more stable is their environment. By stability is meant the ability to return to the original position after any change, instead of being forced into a totally different pattern – and hence predictability.

Unfortunately, we behave as if we knew nothing of the environment and had no conception of its predictability, treating it instead with scant and brutal regard as if it were an idiosyncratic and extremely stupid slave. We seem never to have reflected on the fact that a tropical rain forest supports innumerable insect species and yet is never devastated by them; that its rampant luxuriance is not contingent on our overflying it once a month and bombarding it with insecticides, herbicides, fungicides, and what-have-you. And yet we tremble over our wheatfields and cabbage patches with a desperate battery of synthetic chemicals, in an absurd attempt to impede the operation of the immutable 'law' we have just mentioned – that all ecosystems tend towards stability, therefore diversity and complexity, therefore a growing number of different plant and animal species until a climax or optimal condition is achieved. If we were clever, we would recognize that successful long-term agriculture demands the achievement of an artificial climax, an imitation of the pre-existing ecosystem, so that the level of unwanted species could be controlled by those that did no harm to the crop-plants.

Instead we have put our money on pesticides, which although they have been effective have been so only to a limited and now diminishing extent: according to SCEP, the 34 per cent increase in world food production from 1951 to 1966 required increased investments in nitrogenous fertilizers of 146 per cent and in pesticides of 300 per cent. At the same time they have created a number of serious problems, notably resistance – some 250 pest species are resistant to one group of pesticides or another, while many others require increased applications to keep their populations within manageable proportions – and the promotion of formerly innocuous species to pest proportions, because the predators that formerly kept them down have been destroyed. The spread of DDT and other organochlorines in the environment has resulted in alarming population declines among woodcock, grebes, various birds of prey and seabirds, and in a number of fish species, principally the sea trout. SCEP comments:

The oceans are an ultimate accumulation site of DDT and its residues. As much as 25 per cent of the DDT compounds produced to date may have been transferred to the sea. The amount in the marine biota is estimated to be in the order of less than 0·1 per cent of total production and has already produced a demonstrable impact upon the marine environment ... The decline in productivity of marine food fish and the accumulation of levels of DDT in their tissues which are unacceptable to man can only be accelerated by DDT's continued release to the environment ...

There are half a million man-made chemicals in use today, yet we cannot predict the behaviour or properties of the greater part of them (either singly or in combination) once they are released into the environment. We know, however, that the combined effects of pollution and habitat destruction menace the survival of no fewer than 280 mammal, 350 bird, and 20,000 plant species. To those who regret these losses but greet them with the comment that the survival of *Homo sapiens* is surely more important than that of an eagle or a primrose, we repeat that *Homo sapiens* himself depends on the continued resilience of those ecological networks of which eagles and primroses are integral parts. We do not need to destroy the ecosphere utterly to bring catastrophe upon ourselves: all we have to do is to carry on as we are, clearing forests, 'reclaiming' wetlands, and imposing sufficient quantities of pesticides, radioactive materials, plastics, sewage, and industrial wastes upon our air, water and land systems to make them inhospitable to the species on which their continued stability and integrity depend. Industrial man in the world today is like a bull in a china shop, with the single difference that a bull with half the information about the properties of china as we have about those of ecosystems would probably try and adapt its behaviour to its environment rather than the reverse. By contrast, *Homo sapiens industrialis* is determined that the china shop should adapt to him, and has therefore set himself the goal of reducing it to rubble in the shortest possible time.

FAILURE OF FOOD SUPPLIES

Increases in food production in the undeveloped world have barely kept abreast of population growth. Such increases as there have been are due not to higher productivity but to the opening up of new land for cultivation. Unfortunately this will not be possible for much longer: all the good land in the world is now being farmed, and according to the FAO (2) at present rates of expansion none of the marginal land that is left will be unfarmed by 1985 – indeed some of the land now under cultivation has been so exhausted that it will have to be returned to permanent pasture.

For this reason, FAO's programme to feed the world depends on a programme of intensification, at the heart of which are the new high-yield varieties of wheat and rice. These are highly responsive to inorganic fertilizers and quick-maturing, so that up to ten times present yields can be obtained from them. Unfortunately, they are highly vulnerable to disease, and therefore require increased protection by pesticides, and of course they demand massive inputs of fertilizers (up to 27 times present ones). Not only will these disrupt local ecosystems, thereby jeopardizing long-term productivity, but they force hard-pressed undeveloped nations to rely on the agro-chemical industries of the developed world.

Whatever their virtues and faults, the new genetic hybrids are not intended to solve the world food problem, but only to give us time to devise more permanent and realistic solutions. It is our view, however, that these hybrids are not the best means of doing this, since their use is likely to bring about a reduction in overall diversity, when the clear need is to develop an agriculture diverse enough to have long-term potential. We must beware of those 'experts' who appear to advocate the transformation of the ecosphere into nothing more than a food-factory for man. The concept of a world consisting solely of man and a few favoured food plants is so ludicrously impracticable as to be contemplated seriously only by those who find solace in their own wilful ignorance of the real world of biological diversity.

We in Britain must bear in mind that we depend on imports for half our food, and that we are unlikely to improve on this situation. The 150,000 acres which are lost from agriculture each year are about 70 per cent more productive than the average for all enclosed land (3), while we are already beginning to experience diminishing returns from the use of inorganic fertilizers. In the period 1964–9, applications of phosphates have gone up by 2 per cent, potash by 7 per cent, and nitrogen by 40 per cent (4), yet yields per acre of wheat, barley, lucerne and temporary grass have levelled off and are beginning to decline, while that of permanent grass has risen only slightly and may be levelling off (5). As *per capita* food availability declines throughout the rest of the world, and it appears inevitable it will, we will find it progressively more difficult and expensive to meet our food requirements from abroad. The prospect of severe food shortages within the next thirty years is not so much a fantasy as that of the continued abundance promised us by so many of our politicians.

EXHAUSTION OF RESOURCES

As we have seen, continued exponential growth of consumption of materials and energy is impossible. Present reserves of all but a few metals will be exhausted within 50 years, if consumption rates continue to grow as they are (see Figure 2). Obviously there will be new discoveries and advances in mining technology, but these are likely to provide us with only a limited stay of execution. Synthetics and substitutes are likely to be of little help, since they must be made from materials which themselves are in short supply; while the hoped-for availability of unlimited energy would not be the answer, since the problem is the ratio of useful metal to waste matter (which would have to be disposed of without disrupting ecosystems), not the need for cheap power. Indeed, the availability of unlimited power holds more of a threat than a promise, since energy use is inevitably polluting, and in addition we would ultimately have to face the problem of disposing of an intractable amount of waste heat.

COLLAPSE OF SOCIETY

The developed nations consume such disproportionate amounts of protein, raw materials and fuels that unless they considerably reduce their consumption there is no hope of the undeveloped nations markedly improving their standards of living. This vast differential is a cause of much and growing discontent, made worse by our attempts at cultural uniformity on behalf of an expanding market economy. In the end, we are altering people's aspirations without providing the means for them to be satisfied. In the rush to industrialize we break up communities, so that the controls which formerly regulated behaviour are destroyed before alternatives can be provided. Urban drift is one result of this process, with a consequent rise in anti-social practices,

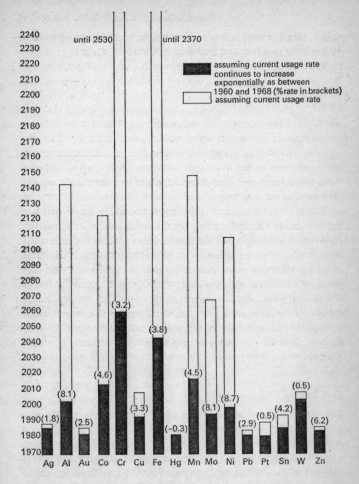

Figure 2   Mineral resources: static and exponential reserves (6).

Ag–silver	Fe–iron	Pt–platinum
Al–aluminium (bauxite)	Hg–mercury	Sn–tin
Au–gold	Mn–manganese	W–tungsten
Co–cobalt	Mo–molybdenum	Zn–zinc
Cr–chromium	Ni–nickel	
Cu–copper	Pb–lead	

crime, delinquency, and so on, which are so costly for society in terms both of money and of well-being.

At the same time, we are sowing the seeds of massive unemployment by increasing the ratio of capital to labour so that the provision of each job becomes ever more expensive. In a world of fast-diminishing resources, we shall quickly come to the point when very great numbers of people will be thrown out of work, when the material compensations of urban life are either no longer available or prohibitively expensive, and consequently when whole sections of society will find good cause to express their considerable discontent in ways likely to be anything but pleasant for their fellows.

It is worth bearing in mind that the barriers between us and epidemics are not so strong as is commonly supposed. Not only is it increasingly difficult to control the vectors of disease, but it is more than probable that urban populations are being insidiously weakened by overall pollution levels, even when they are not high enough to be incriminated in any one illness. At the same time international mobility speeds the spread of disease. With this background, and at a time of widespread public demoralization, the collapse of vital social services such as power and sanitation could easily provoke a series of epidemics – and we cannot say with confidence that we would be able to cope with them.

At times of great distress and social chaos, it is more than probable that governments will fall into the hands of reckless and unscrupulous elements, who will not hesitate to threaten neighbouring governments with attack if they feel that they can wrest from them a larger share of the world's vanishing resources. Since a growing number of countries (an estimated 36 by 1980) will have nuclear power stations, and therefore sources of plutonium for nuclear warheads, the likelihood of a whole series of local (if not global) nuclear engagements is greatly increased.

CONCLUSION

A fuller discussion of ecosystems and their disruption, of social systems and their disruption, of population and food supply, and of resources and their depletion, can be found in Appendices A, B, C and D, respectively. There will be those who regard these accounts of the consequences of trying to accommodate present growth rates as fanciful. But the imaginative leap from the available scientific information to such predictions is negligible, compared with that required for those alternative predictions, laughably considered 'optimistic', of a world of 10,000 to 15,000 million people, all with the same material standard of living as the US, on a concrete replica of this planet, the only moving parts being their machines and possibly themselves. Faced with inevitable change, we have to make decisions, and we must make these decisions *soberly* in the light of the best information, and not as if we were caricatures of the archetypal mad scientist.

By now it should be clear that the main problems of the environment do not arise from temporary and accidental malfunctions of existing economic and social systems. On the contrary, they are the warning signs of a profound incompatibility between deeply rooted beliefs in continuous growth and the dawning recognition of the earth as a space ship, limited in its resources and vulnerable to thoughtless mishandling. The nature of our response to these symptoms is crucial. If we refuse to recognize the cause of our trouble the result can only be increasing disillusion and growing strain upon the fragile institutions that maintain external peace and internal social cohesion. If, on the other hand, we can respond to this unprecedented challenge with informed and constructive action the rewards will be as great as the penalties for failure.

We are sufficiently aware of 'political reality' to appreciate that many of the proposals we will make in the next chapter will be considered impracticable. However, we believe that if a strategy for survival is to have any chance of success, the solutions must be formulated in the light of the problems and not from a timorous and superficial understanding of what may or may not be immediately feasible. If we plan remedial action with our eyes on political rather than ecological reality, then very reasonably, very practicably, and very surely, we will muddle our way to extinction.

A measure of political reality is that government has yet to acknowledge the impending crisis. This is to some extent because it has given itself no machinery for looking at energy, resources, food, environmental disruption and social disruption as a whole, as part of a general, global pattern, preferring instead to deal with its many aspects as if they were self-contained analytical units. Lord Rothschild's Central Policy Review Staff in the Cabinet Office, which is the only body in government which might remedy the situation, appears not to think it worthwhile: at the moment at least, they are undertaking 'no specific studies on the environment that would require an environmentalist or ecologist'. There is a strong element of positive feedback here, in that there can be no appreciation of our predicament unless we view it in totality, and yet government can see no cause to do so unless it can be shown that such a predicament exists.

Possibly because government sees the world in fragments and not as a totality, it is difficult to detect in its actions or words any coherent general policy, although both major political parties appear to be mesmerized by two dominating notions: that economic expansion is essential for survival and is the best possible index of progress and well-being; and that unless solutions can be devised that do not threaten this notion, then the problems should not be regarded as existing. Unfortunately, government has an increasingly powerful incentive for continued expansion in the tendency for economic growth to create the need for more economic growth. This it does in six ways:

Firstly, the introduction of technological devices, i.e. the growth of the technosphere, can only occur to the detriment of the ecosphere, which means that it leads to the destruction of natural controls which must then be replaced by further technological ones. It is in this way that pesticides and artificial fertilizers create the need for yet more pesticides and artificial fertilizers.

Secondly, for various reasons, industrial growth, particularly in its earlier phases, promotes population growth. Even in its later phases, this can still occur at a high rate (0·5 per cent in the UK). Jobs must constantly be created for the additional people

– not just any job, but those that are judged acceptable in terms of current values. This basically means that the capital outlay per person employed must be maintained, otherwise the level of 'productivity' per man will fall, which is a determinant of both the 'viability' of economic enterprise and of the 'standard of living'.

Thirdly, no government can hope to survive widespread and protracted unemployment, and without changing the basis of our industrial society, the only way government can prevent it is by stimulating economic growth.

Fourthly, business enterprises, whether state-owned or privately owned, tend to become self-perpetuating, which means that they require surpluses for further investment. This favours continued growth.

Fifthly, the success of a government and its ability to obtain support is to a large extent assessed in terms of its ability to increase the 'standard of living' as measured by *per capita* gross national product (GNP).

Finally, confidence in the economy, which is basically a function of its ability to grow, must be maintained to ensure a healthy state of the stock market. Were confidence to fail, stock values would crash, drastically reducing the availability of capital for investment and hence further growth, which would lead to further unemployment. This would result in a further fall in stock-market values and hence give rise to a positive-feedback chain-reaction, which under the existing order might well lead to social collapse.

For all these reasons, we can expect our government (whether Conservative or Labour) to encourage further increases in GNP regardless of the consequences, which in any case tame 'experts' can be found to play down. It will curb growth only when public opinion demands such a move, in which case it will be politically expedient, and when a method is found for doing so without creating unemployment or excessive pressure on capital. We believe this is possible only within the framework of a fully integrated plan.

The emphasis must be on integration. If we develop relatively clean technologies but do not end economic growths then sooner or later we will find ourselves with as great a pollution problem as before but without the means of tackling it. If we stabilize our economies and husband our non-renewable resources without stabilizing our populations we will find we are no longer able to feed ourselves. As Forrester (1) and Meadows (7) convincingly make clear, daunting though an integrated programme may be, a piecemeal approach will cause more problems than it solves.

Our task is to create a society which is sustainable and which will give the fullest possible satisfaction to its members. Such a society by definition would depend not on expansion but on stability. This does not mean to say that it would be stagnant – indeed it could well afford more variety than does the state of uniformity at present being imposed by the pursuit of technological efficiency. We believe that the stable society, the achievement of which we shall discuss in the next chapter, as well as removing the sword of Damocles which hangs over the heads of future generations, is much more likely than the present one to bring the peace and fulfilment which hitherto have been regarded, sadly, as utopian.

2   Towards the Stable Society: Strategy for Change

The principal conditions of a stable society – one that to all intents and purposes can be sustained indefinitely while giving optimum satisfaction to its members – are: (1) minimum disruption of ecological processes; (2) maximum conservation of materials and energy – or an economy of stock rather than flow; (3) a population in which recruitment equals loss; and (4) a social system in which the individual can enjoy, rather than feel restricted by, the first three conditions.

The achievement of these four conditions will require controlled and well-orchestrated change on numerous fronts, and this change will probably occur through seven operations:

1. a control operation whereby environmental disruption is reduced as much as possible by technical means;
2. a freeze operation, in which present trends are halted;
3. asystemic substitution, by which the most dangerous components of these trends are replaced by technological substitutes, whose effect is less deleterious in the short term, but over the long term will be increasingly ineffective;
4. systemic substitution, by which these technological substitutes are replaced by 'natural' or self-regulating ones, i.e. those which either replicate or employ without undue disturbance the normal processes of the ecosphere, and are therefore likely to be sustainable over very long periods of time;
5. the invention, promotion and application of alternative technologies which are energy and materials conservative, and which because they are designed for relatively 'closed' economic communities are likely to disrupt ecological processes only minimally (e.g. intermediate technology);
6. decentralization of polity and economy at all levels, and the

formation of communities small enough to be reasonably self-regulating and self-supporting; and

7. education for such communities.

As we shall see when we examine how our four conditions might be achieved, some changes will involve only a few of these operations, in others a number of the operations will be carried out almost simultaneously, and in others one will start well before another has ended. The usefulness of the operation-concept is simply to clarify the orchestration of change.

In putting forward these proposals we are aware that hasty or disordered change is highly disruptive and ultimately self-defeating; but we are also mindful of how the time-scale imposed on any proposal for a remedial course of action has been much abbreviated by the dynamic of exponential growth (of population, resource depletion and pollution) and by the scarcely perceived scale and intensity of our disruption of the ecological processes on which we and all other life-forms depend. Within these limitations, therefore, we have taken care to devise and synchronize our programme so as to minimize both unemployment and capital outlay. We believe it possible to change from an expansionist society to a stable society without loss of jobs or an increase in real expenditure. Inevitably, however, there will be considerable changes, both of geography and function, in job availability and the requirements for capital inputs – and these may set up immense counter-productive social pressures. Yet given the careful and sensitive conception and implementation of a totally integrated programme these should be minimized, and an open style of government should inspire the trust and co-operation of the general public so essential for the success of this enterprise.

One further point should be made before we consider in more detail the various changes required. As each of the many socio-economic components or variables of industrial society are changed or replaced, so various pressure-points will be set up. It is easy to imagine, for example, a situation in which 25 per cent of the socio-economic variables are designed for a stable society and therefore by definition are ill-suited to one of expansion. This situation may create more problems than it solves. When we reach the point at which 50 per cent of the variables are adapted to stability and the other 50 per cent to expansion, the difficulties and tensions are likely to be enormous, but thereafter each change and replacement will assist further change and replacement, and the moulding of a sustainable, satisfying society should be that much easier. It is difficult for the human mind to imagine the temporal sequence of complex change, and no doubt impossible for it to visualize the precise interactions of the various components. While bearing in mind the folly of expecting computers to do our thinking for us, we believe they have an important role to play in demonstrating the consequences throughout social and ecological systems of a great number of changes over a given period of time.

MINIMIZING THE DISRUPTION OF ECOLOGICAL PROCESSES

Ecological processes can be disrupted by introducing into them either substances that are foreign to them or the correct ones in the wrong quantities. It follows therefore that the most common method of pollution 'control', namely dispersal, is not control at all, but a more or less useful way of playing for time. Refuse disposal by dumping solves the immediate problem of the householder, but as dumping sites are used up it creates progressively less soluble problems for society at large; smoke-less fuels are invaluable signs of progress for the citizens of London or Sheffield, but the air pollution from their manufacture brings misery and ill-health to the people near the plants where they are produced; in many cases the dispersal of pollutants through tall chimneys merely alters the proportion of pollution, so that instead of a few receiving much, many receive some; and lastly, in estuarine and coastal waters – crucial areas for fisheries – nutrients from sewage and agricultural run-off in modest quantities probably increase productivity, but in excess are as harmful as organochlorines and heavy metals.

Thus dispersal can be only a temporary expedient. Pollution control proper must consist of the recycling of materials, or the introduction of practices which are so akin to natural processes as not to be harmful. The long-term object of these pollution-control procedures is to minimize our dependence on technology as a regulator of the ecological cycles on which we depend, and to return as much as possible to the natural mechanisms of the ecosphere, since in all but the short term they are much more efficient and reliable. In the light of these remarks then, let us consider some contemporary pollution problems and how they might be solved.

Pesticides

There is no way of controlling the disruption caused by pesticides save by using less, and progress towards this end will probably require three operations: freeze, asystemic substitution, and systemic substitution. The freeze operation consists of the ending of any further commitment to pesticides, particularly the persistent organochlorines. For the developed countries this is a relatively simple procedure, and already the use of Dieldrin, DDT, and so on, is beginning to decline. For the undeveloped countries, however, it would be impossible without an undertaking from the developed ones to subsidize the supply of much more expensive substitutes. In the malaria control programme, for example, the replacement of DDT by malathion or propoxur would raise the cost of spraying operations from US $60 million a year to $184 million and $510 million respectively (1).

Once such an undertaking is given, the undeveloped countries could proceed to the second operation. (There is no conceivable reason why the developed ones should not formally do so now.) This consists of the progressive substitution of non-persistent pesticides (organophosphates, carbamates, etc.) for the organochlorines. The third operation, the substitution of natural controls for pesticides in general, could follow soon after. Two important points should be borne in mind: (a) it is most unlikely that the third stage could ever be complete – we will probably have to rely on the precision use of pesticides for some considerable time as part of a programme of integrated control; and (b) the second and third operations would proceed in harness until all countries had fully integrated pest control programmes. The drawback with integrated control (the combination of biological control, mechanical control, crop-species diversity and the precise use of species-specific pesticides) is that as yet we do not know enough about it, so that a full-scale research programme is urgently required. The agro-chemical industries should be encouraged to invest in integrated control programmes though plainly, since the profits cannot be so great as from chemical control, research will need public finance – as will the training of integrated control advisory teams to assist farmers, particularly in the undeveloped countries. Such an investment, however, will appear modest once integrated control is fully operational, in comparison with the vast sums of money currently being spent annually on pesticides. A typical operational procedure for the transfer from chemical to integrated control might be as follows: organochlorines phased out, substitute pesticides phased in; in some cultivations these substitutes would be phased out almost immediately, to be replaced by integrated control; in others the time-table would be somewhat longer, depending on our understanding of the relevant agro-ecological processes and the availability of trained personnel.

Fertilizers

While on many occasions the use of inorganic fertilizers is valuable, their overuse leads to two intractable problems: the pollution of freshwater systems by run-off, and diminishing returns due to the slow but inevitable impoverishment of the soil (see Appendix C on food supply). Again the solution will come through three operations: freeze, asystemic substitution, and systemic substitution. The first operation requires there to be no further increment in the application of inorganic fertilizers, and hence the removal of subsidies for them. Again this is relatively easy for the developed countries (although there may be some drop in yield per acre), but next to impossible for the undeveloped countries, which are now being introduced to the new genetic hybrids of rice and wheat. Since the remarkable responsiveness of these hybrids is contingent on massive fertilizer inputs (up to 27 times present ones), the undeveloped world is faced with an unenviable choice: either to keep alive its expanding population over the next ten years at the price of considerable damage to soil structure and long-term fertility; or to improve soil structure so that a good proportion of the population can be fed indefinitely, but in the knowledge that the population will probably be reduced to that proportion by such natural processes as famine and epidemic. In the long term, of course, the solution lies in population control; but in the intervening period there seems to be no alternative to concentrating on agricultural methods that are sustainable even at the expense of immediate productivity. The consequences of not doing so are likely to be much worse than any failure to take full advantage of the new hybrids. In the meantime, an emergency food-supply must be created by the developed prime-producers (USA, USSR, Canada, Australia, New Zealand) so that as much as possible of any short-fall can be met during this difficult period.

The second operation involves the gradual substitution of organic manures for inorganic fertilizers – though occasionally the latter will be used to supplement the former – and the return to such practices as rotation and leys; this would merge into the third operation: the adoption of highly diversified farming practices in place of monocultures. It is necessary to emphasize that this is not simply a return to traditional good husbandry: it is much more a change from flow fertility (whereby nutrients are imported from outside the agro-ecosystem, a proportion being utilized by food-plants, but with a large proportion leaving the agro-ecosystem in the form of run-off, etc.) to cyclic fertility (in which nutrients in the soil are used and then returned to it in as closed a cycle as possible). The great advantage of nutrients in organic form is that the soil appears much better adapted to them. The nitrogen in humus, for example, is only 0·5 per cent inorganic, the rest being in the form of rotting vegetation, decomposing insects and other animals, and animal manure. A high proportion of organic matter is essential for the soil to be easily workable over long periods (thus extending the period in which cultivations are timely), for it to retain water well without becoming saturated, for the retention of nutrients so that they remain available to plants until they are taken up by them (thus

reducing wastage), and for the provision of the optimum environment for the micro-organisms so vital for long-term fertility.
The rotation of leguminous plants and of grass grazed by animals
are the most effective ways of adding organic matter to the soil,
while at the same time allowing livestock to select their own
food in the open has the double advantage that they are bred
with a healthy fat-structure and their wastes enrich the soil
instead of polluting waterways or overloading sewage systems.
By diversifying farming in these and other ways we are taking
advantage of the immense growth of knowledge about agricultural ecology, which plainly will increase with additional research.

Domestic sewage

The volume of sewage is directly proportional to population
numbers and can only be stabilized or reduced by stabilizing or
reducing the population. However, sewage can and should be
disposed of much more efficiently. It is absurd that such valuable nutrients should be allowed to pollute fresh and coastal
waters, or that society should be put to the expense of disposing
of them in areas where they cannot be effectively utilized.
Unfortunately, in developed countries, their disposal as agricultural fertilizer is not generally feasible, largely for two
reasons: (a) they are contaminated by industrial wastes; (b)
transportation costs are too high. Both difficulties can be overcome – in the first case by ensuring that there is no (or negligible)
admixture of industrial to domestic effluents, which depends on
better industrial pollution control (see below); and in the second
case by decentralizing so that there is an improved mix of rural
and urban activities. This will be explored in the section on
social systems. In undeveloped countries, the problem of domestic sewage could be overcome by the provision of aid to pay for
sewage plants that yield purified water and usable sludge.

Industrial wastes

Reduction of industrial effluent should proceed by two operations: a control operation, and an alternative (materials and

energy conservative) technology operation. We have already suggested that the key to pollution control is not dispersal but recycling, and since recycling is a most important element in resource management it will be discussed in the section on stock economics. The alternative technology operation will be considered in the section on social systems.

CONVERSION TO AN ECONOMY OF STOCK

The transfer from flow to stock economics can be considered under two headings: resource management and social accounting.

Resource management

It is essential that the throughput of raw materials be minimized both to conserve non-renewable resources and to cut down pollution. Since industry must have an economic incentive to be conservative of materials and energy and to recycle as much as possible, we propose a number of fiscal measures to these ends:

(*a*) A raw materials tax. This would be proportionate to the availability of the raw material in question, and would be designed to enable our reserves to last over an arbitrary period of time, the longer the better, on the principle that during this time our dependence on this raw material would be reduced. This tax would penalize resource-intensive industries and favour employment-intensive ones. Like (*b*) below, it would also penalize short-lived products.

(*b*) An amortization tax. This would be proportionate to the estimated life of the product, e.g. it would be 100 per cent for products designed to last no more than a year, and would then be progressively reduced to zero per cent for those designed to last 100+ years. Obviously this would penalize short-lived products, especially disposable ones, thereby reducing resource utilization and pollution, particularly the solid-waste problem. Plastics, for example, which are so remarkable for their durability, would be used only in products where this quality is valued, and not for

single-trip purposes. This tax would also encourage craftsmanship and employment-intensive industry.

The raw materials tax would obviously encourage recycling,
and we can see how it might work if we consider such a vital
resource as water. The growing conflict between farmers, con-
servationists and the water boards is evidence enough that
demand for water is conflicting with other, no less important,
values. At the moment, the water boards have no alternative but
to fulfil their statutory obligation to meet demand, and accordingly valley after valley comes under the threat of drowning.
Clearly, unless we consider dry land an obstacle to progress,
demand must be stabilized, and since demand is a function of
population numbers × *per capita* consumption, both must be
stabilized, if not reduced (and we have seen that for other
reasons they must be reduced). To this end therefore, while a
given minimum can be supplied to each person free-of-charge,
any amount above that minimum should be made increasingly
expensive. As far as industry is concerned, the net effect would
be to encourage the installation of closed-circuit systems for
water; total demand would be reduced, and there would be less
pressure on lowland river systems.

Despite the stimulus of a raw materials tax, however, it is
likely that there would be a number of serious pollutants which
it would be uneconomic to recycle, and still others for which
recycling would be technically impossible. One thinks in particular of the radioactive wastes from nuclear power stations.
Furthermore, recycling cannot do everything: there will always
be a non-recoverable minimum, which as now will have to be
disposed of as safely as possible. This limitation can be made
clear if we postulate a 3 per cent growth rate, and the introduction of pollution controls which reduce pollution by 80 per cent
throughout – it would then take only 52 years to bring us back
where we started from, with the original amount of pollution
but with a much greater problem of reducing it any further; if
we had a 6 per cent growth rate, we would reach this position
in a mere 26 years. It is also worth mentioning that recycling
consumes energy and is therefore polluting, so that it is necessary
to develop recycling procedures which are energy conservative.

The problem of uneconomic recycling can be resolved by the granting of incentives by government. Indeed, in the short term, the entire recycling industry should be encouraged to expand, even though we know that in the long term industrial expansion is self-defeating. This brings us to the intractable problem of the disposal of the undisposable, which can only be resolved by the termination of industrial growth and the reduction of energy demand. Again fiscal measures will be supremely important, and we propose one in particular:

(c) A power tax. This would penalize power-intensive processes and hence those causing considerable pollution. Since machinery requires more power than people, it would at the same time favour the employment intensification of industry, i.e. create jobs. It would also penalize the manufacture of short-lived products. In addition to this tax, there should be financial incentives for the development and installation of total energy systems, a matter to which we shall return in the section on social systems.

Finally, industrial pollution can also be reduced by materials substitution. The substitution of synthetic compounds for naturally occurring compounds has created serious environmental damage since in some cases the synthetics can be broken down only with difficulty and in others not at all. The usage rate of these synthetics has increased immensely at the expense of the natural products, as can be seen from the following examples (2):

(a) In the US, *per capita* consumption of synthetic detergents increased by 300 per cent between 1962 and 1968. They have largely replaced soap products, *per capita* consumption of which fell by 71 per cent between 1944 and 1964.

(b) Synthetic fibres are rapidly replacing cotton, wool, silk and other natural fibres. In the US, *per capita* consumption of cotton fell by 33 per cent between 1950 and 1968.

(c) The production of plastics and synthetic resins in the US rose by 300 per cent between 1958 and 1968. They have largely replaced wood and paper products.

All of these processes consume the non-renewable fossil fuels, and their manufacture requires considerable inputs of energy.

On the face of it, therefore, a counter-substitution of naturally occurring products would much reduce environmental disruption. However, it is possible that such a change-over, while it would certainly reduce disruption at one end, might dangerously increase it at the other. For example, many more acres would have to be put under cotton, thus increasing demand for pesticides, more land would have to be cleared and put under forest monocultures, and so on. This problem can be solved only by reducing total consumption.

Genetic resources

Before leaving the subject of resources, it is appropriate that we consider the world's diminishing stock of genetic resources. Genetic diversity is essential for the security of our food supply, since it is the *sine qua non* of plant breeding and introduction. The greater the number of varieties, the greater the opportunities for developing new hybrids with resistance to different types of pests and diseases, and to extremes of climate. It is important that new hybrids be continually developed since resistance to a particular disease is never a permanent quality. The number of plant varieties to be found in nature is infinitely greater than the number we could create artificially. Most of them are to be found in the undeveloped countries either as traditional domesticated plants or as wild plants in habitats relatively unaltered by man. There is a real danger that the former will be replaced by contemporary high-yield varieties, while the latter will disappear when their habitats are destroyed. An FAO conference in 1967 concluded that the plant-gene pool has diminished dangerously, for all over the world centres of diversity, our gene banks as it were, are disappearing, and with them our chance of maintaining productivity in food (3).

Such centres – areas of wilderness – are often destroyed because their importance is not understood. Because they seem less productive than fields of waving corn, or because they are not accessible or attractive to tourists, they are considered in need of 'improvement' or development, or simply as suitable dumping grounds for the detritus of civilization. This is particularly true of wetlands – estuaries and marshes – where pollution, dredging, draining and filling are looked on almost with equanimity, certainly with scant regard for what is being lost. Yet the complex of living and decomposing grasses, and of phytoplankton, characteristic of wetlands, supports vast numbers of fish and birds and makes it one of the world's most productive ecosystems. Estuaries are the spawning grounds of very many fish and shellfish and form the base of the food-chain of some 60 per cent of our entire marine harvest. Should they go we can expect a substantial drop in productivity.

It is vital to the future well-being of man that wilderness areas and wetlands be conserved at all costs. This cannot be a matter simply of taking seed and storing it, since to be valuable genetic stock must continue to be subject to normal environmental pressures, and besides we have scarcely any idea of what plants we shall find useful in the future. For these reasons we must not only conserve large areas of natural habitat, we must also draw upon the knowledge and experience of the hunter-gatherers and hunter-farmers who gain their livelihood from them.

We therefore have recommended to the UN Human Environment Conference that (4):

1. certain wilderness areas of tropical rain forest, tropical scrub forest, and arctic tundra be declared inviolate, these being the least understood and most fragile biomes;
2. the hunter-gatherers and hunter-farmers within these areas be given title to their lands (i.e. those lands in which traditionally they have gained their living) and be allowed to live there without pressure of any kind;
3. severe restrictions be placed on entry to these areas by anyone who does not live there permanently (while allowing the indigenes free movement);
4. sovereignty over the areas remain with the countries in which they lie; who should also be responsible for the policing of their boundaries;
5. funds for administration of these areas and payments in lieu of exploitation (to the host country) be collected from UN members in proportion to their GNP;
6. an international body be appointed as an outcome of the

Stockholm Human Environment Conference to supervise an ecological programme of research, the results of which should be freely available to participating countries.

Social accounting

By the introduction of monetary incentives and disincentives it is possible to put a premium on durability and a penalty on disposability, thereby reducing the throughput of materials and energy so that resources are conserved and pollution reduced. But another important way of reducing pollution and enhancing amenity is by the provision of a more equitable social accounting system, reinforced by anti-disamenity legislation. Social accounting procedures must be used not just to weigh up the merits of alternative development proposals, but also to determine whether or not society actually wants such development. Naturally, present procedures require improvement: for example, in calculating 'revealed preference' (the values of individuals and communities as 'revealed' to economists by the amount people are willing and/or can afford to pay for or against a given development), imagination, sensitivity and common sense are required in order to avoid the imposition on poor neighbourhoods or sparsely inhabited countryside of nuclear power stations, reservoirs, motorways, airports, and the like; and in calculating the 'social time preference rate' (an indication of society's regard for the future) for a given project, a very low discount should be given, since it is easier to do than undo, and we must assume that unless we botch things completely many more generations will follow us who will not thank us for exhausting resources or blighting the landscape.

The social costs of any given development should be paid by those who propose or perpetrate it – 'the polluter must pay' is a principle that must guide our costing procedures. Furthermore, accounting decisions should be made in the light of stock economics: in other words, we must judge the health of our economy not by flow or throughput, since this inevitably leads to waste, resource depletion and environmental disruption, but by the distribution, quality and variety of the stock. At the moment, as Kenneth Boulding has pointed out (5), 'the success of the economy is measured by the amount of throughput derived in part from reservoirs of raw materials, processed by "factors of production", and passed on in part as output to the sink of pollution reservoirs. The Gross National Product (GNP) roughly measures this throughout'. Yet, both the reservoirs of raw materials and the reservoirs for pollution are limited and finite, so that ultimately the throughput from the one to the other must be detrimental to our well-being, and must therefore be not only minimized but regarded as a cost rather than a benefit. For this reason Boulding has suggested that GNP be considered a measure of gross national cost, and that we devote ourselves to its minimization, maximizing instead the quality of our stock. He writes:

When we have developed the economy of the spaceship earth in which man will persist in equilibrium with his environment, the notion of the GNP will simply disintegrate. We will be less concerned with income-flow concepts and more with capital-stock concepts. Then technological changes that result in the maintenance of the total stock with *less* throughput (less production and consumption) will be a clear gain.

We must come to assess our standard of living not by calculating the value of all the air-conditioners we have made and sold, but by the freshness of the air; not by the value of the antibiotics, hormones, feedstuff and broiler-houses, and the cost of disposing of their wastes, all of which put so heavy a price on poultry production today, but by the flavour and nutritional quality of the chickens themselves; and so on. In other words, accepted value must reflect real value, just as accepted cost must reflect real cost.

It is evident, however, that in a society such as ours, which to a large extent ignores the long-term consequences of its actions, there is a substantial differential between accepted cost and real cost. An industrial town, for example, whose citizens and factories pollute the air and water systems around it and who feed themselves from a number of increasingly intensive monocultures, not only has no way of measuring the satisfactions or otherwise afforded by its life-style, nor of equitably distributing the costs imposed by one polluter on another, but no way either of assessing ecological costs, some of which will have to be paid by generation 1, others by generations 2, 3, 4, etc., and still others by people elsewhere, with whom in every other respect there might be no contact. Thus its agricultural practices might provide cheap and plentiful food for one generation and stimulate its agro-chemical industries, but may so impoverish the soil and disrupt the agro-ecosystem that the next generation will have to import more food, or, failing this, resort to still riskier expedients, thereby serious compromising the food supply of the following generation; or the wastes of one generation might affect the health of the next, or its marine food supply, or so increase the mutation rate that future generations receive an unlooked-for genetic burden. The extent to which we are simplifying ecosystems and destroying natural controls so that we are forced to provide technological substitutes is a real cost against society and should be accounted as one. At the moment, however, we merely add up the value of mining operations, factories and so on, and that of cleaning up the mess whenever we attempt to do so, and conclude that we have never been better off.

Since the full costs of any action anywhere in the world must be borne by someone, somewhere, sometime, it is important that our accounting system makes provision for this. We accept, however, that ecological processes are so complex, and can spread so far in space and time, that this will be exceptionally difficult. Nonetheless, given the truism that a satisfactory accounting system is one which supports and helps perpetuate the social system from which it derives, we must attempt to devise one which is fitted to a society based on a sober assessment of ecological reality and not on the anthropocentric pipe-dream that we can do what we will to all species, not excepting, it seems, future generations of our own. It is worth recalling Professor Commoner's dictum that since economics is the science of the distribution of resources, all of which are derived from the ecosphere, it is foolish to perpetuate an economic system which destroys it. Ideally (and as befits the etymology of the two words), ecology and economics should not be in conflict: ecology should provide the approach, the framework for an understanding of the interrelationships of social and environmental systems; and economics should provide the means of quantifying those inter-relationships in the light of such an understanding, so that decisions on alternative courses of action can be made without undue difficulty.

One of our long-term goals, therefore, must be to unite economics and ecology. The specific measures we have proposed are, we believe, necessary steps in this direction, albeit crude ones. A raw materials tax, an amortization tax, a power tax, revised methods of calculating revealed preference, social time preference rate, and so on, with legislative provision for their enforcement, a set of air, water and land quality standards enforceable at law and linked with a grant-incentive programme – these and other measures will have to be introduced at an early stage. Naturally, the full force of such measures could not be allowed to operate immediately: they would have to be carefully graded so as to be effective without causing unacceptable degrees of social disturbance. Plainly the social consequences will be great, and these will be considered in the section on social systems. The key to success is likely to be careful synchroniza-tion, and this too will be considered in a separate section.

STABILIZING THE POPULATION

We have seen already that, however slight the growth rate, a population cannot grow indefinitely. It follows, therefore, that at some point it must stabilize of its own volition, or else be cut down by some 'natural' mechanism – famine, epidemic, war, or whatever. Since no sane society would choose the latter course, it must choose to stabilize. To do this it must have some idea of its optimum size, or at least of its maximum sustainable size, since the former figure is extremely difficult to work out.

The two main variables affected by population numbers, as opposed to *per capita* consumption, are the extent to which the emotional needs and social aspirations of the community can be met (i.e. the complex of satisfactions which has come to be known as the quality of life), and the community's ability to feed itself. It is clear that our population is too high to permit the optimization of many social and ecological requirements. It would be advantageous, for instance, for many reasons to increase substantially the amount of forests in this country, which we cannot do, mainly because land is required for other purposes, such as agriculture, roads, housing, etc. In the final analysis, the maximum sustainable population is that which we are capable of feeding, which depends on the single ecological concept of the carrying capacity of the land.

Carrying capacity is usually defined as the amount of solar energy potentially available to man via food-plants in a given area. This definition must be accompanied by a caveat to the effect that, if carrying capacity is considered in terms of energetics alone, a number of essential ecological and nutritional variables are in danger of exclusion. For example, it would be easy to assume that land used for a combination of purposes (mixed farming, woodland, etc.) would be better employed and could support a larger population if it were exclusively given over to the intensive production of food-plants high in calories (e.g. wheat). We know, however, that protein and the other nutrients are no less vital to us than calories, while there is evidence that we are more likely to get the proper nutritional components from meat if it comes to us from free-living animals. This requirement alone demands a certain diversity, both of species and habitat, and we have found too (see Appendix on ecosystems) that diversity is essential if fertility and stability are to be maintained over the long term.

As we have seen Britain supports a population well in excess of the carrying capacity of the land owing to its ability to import large amounts of food, especially the cheap protein required to feed our poultry and pigs. As world population grows, and with it global agricultural demand, so will it be increasingly difficult for us to find countries with exportable surpluses, surpluses which in any case will become progressively more expensive. Unless we are willing (and able) to perpetuate an even greater inequality of distribution than exists today, Britain must be self-supporting. We have stated already our belief that on the evidence available it is unlikely that there will be any significant increase in yield per acre, so that there is no other course open to us but to reduce our numbers before we stabilize. Since we appear capable of supporting no more than half our present population, the figure we should aim for over the next 150 to 200 years can be no greater than 30 million, and in order to protect it from resource fluctuation probably less.

Not every country is in such a difficult position as Britain. A few will be able to stabilize at or relatively near present levels. But taking world population as a whole, and using daily *per capita* protein intake as the key variable in assessing carrying capacity, we believe the optimum population for the world is unlikely to be above 3,500 million and is probably a good deal less. This figure rests on three assumptions: (*a*) that the average daily *per capita* requirement of protein is 65 grams (6); (*b*) that present agricultural production *per capita* can be sustained indefinitely; and (*c*) that there is absolutely equitable distribution, no country enjoying a greater daily *per capita* protein intake than any other – which compared with today's conditions is absurdly utopian. Utopian though they may be, unless these assumptions are realized, we are faced either with the task of reducing world population still further until it is well below the optimum, or with condoning inequalities grosser and more unjust than those which we in the developed countries foster at present.

While they cannot grow indefinitely, populations can remain above the optimum – indeed above the sustainable maximum – for some time. The fact that the global population, including that of Britain, is above both levels, means only that our numbers are preventing the optimization of other values. It means that while most people receive the bare minimum of calories necessary for survival a large proportion are deprived of the nutrients (especially protein) essential for intellectual development. They are alive, but unable to realize their full potential – which is the grossest possible waste of human resources. An optimum population, therefore, may be defined as one that can be sustained indefinitely and at a level at which the other values of its members are optimized – and the fact that we are above this level does not justify despair, but does justify a great sense of urgency in working towards our long-term goal of the optimum. For it is obvious that given the dynamic of population growth, even if all nations today determined to stabilize their populations, numbers would continue to rise for some considerable time. Indeed the Population Council has calculated (Annual Report 1970) that

. . . if the replacement-sized family is realized for the world as a whole by the end of this century – itself an unlikely event – the world's population will then be 60 per cent larger or about 5·8 billion, and due to the resulting age structure it will not stop growing until near the end of the next century, at which time it will be about 8·2 billion (8,200 million) or about 225 per cent the present size. If replacement is achieved in the developed world by 2000 and in the developing world by 2040, then the world's population will stabilise at nearly 15·5 billion (15,500 million) about a century hence, or well over four times the present size.

Clearly we must go all out for the 'unlikely event' of achieving the replacement-sized family (an average of about two children per couple) *throughout the world by the end of this century*, if our children are not to suffer the catastrophes we seek to avoid.

Our task is to end population growth by lowering the rate of recruitment so that it equals the rate of loss. A few countries will then be able to stabilize, to maintain that ratio; most others, however, will have to *reduce* their populations slowly to a level at which it is sensible to stabilize. Stated baldly, the task seems impossible; but if we start now, and the exercise is spread over a sufficiently long period of time, then we believe that it is within our capabilities. The difficulties are enormous, but they are surmountable.

First, governments must acknowledge the problem and declare their commitment to ending population growth; this commitment should also include an end to immigration. Secondly, they must set up national population services with a fourfold brief:

1. To publicize as widely and vigorously as possible the relationship between population, food supply, quality of life, resource depletion, etc., and the great need for couples to have

no more than two children. The finest talents in advertising should be recruited for this, and the broad aim should be to inculcate a socially more responsible attitude to child-rearing. For example, the notion (derived largely from the popular women's magazines) that childless couples should be objects of pity rather than esteem should be sharply challenged; and of course there are many similar notions to be disputed.

2. To provide at local and national levels free contraception advice and information on other services such as abortion and sterilization.

3. To provide a comprehensive domiciliary service, and to provide contraceptives free of charge, free sterilization, and abortion on demand.

4. To commission, finance, and co-ordinate research not only on demographic techniques and contraceptive technology, but also on the subtle cultural controls necessary for the harmonious maintenance of stability. We know so little about the dynamics of human populations that we cannot say whether the first three measures would be sufficient. It is self-evident that if couples still wanted families larger than the replacement-size no amount of free contraception would make any difference. However, because we know so little about population control, it would be difficult for us to devise any of the socio-economic restraints which on the face of it are likely to be more effective, but which many people fear might be unduly repressive. For this reason, we would be wise to rely on the first three measures for the next twenty years or so. We then may find they are enough – but if they aren't, we must hope that intensive research during this period will be rewarded with a set of socio-economic restraints that are both *effective* and *humane*. These will then constitute the third stage, and should also provide the tools for the fourth stage – that of persuading the public to have average family sizes of slightly *less* than replacement size, so that total population can be greatly reduced. If we achieve a decline rate of 0·5 per cent per year, the same as Britain's rate of growth today, there should be no imbalance of population structure, as the dependency ratio would be exactly the same as that of contemporary Britain. Only the make-up of dependency would be different: instead of

there being more children than old people, it would be the other way round. The time-scale for such an operation is long of course, and this will be suggested in the section on orchestration.

CREATING A NEW SOCIAL SYSTEM

Possibly the most radical change we propose in the creation of a new social system is decentralization. We do so not because we are sunk in nostalgia for a mythical little England of fêtes, olde worlde pubs, and perpetual conversations over garden fences, but for four much more fundamental reasons.

1. While there is good evidence that human societies can happily remain stable for long periods, there is no doubt that the long transitional stage that we and our children must go through will impose a heavy burden on our moral courage and will require great restraint. Legislation and the operations of police forces and the courts will be necessary to reinforce this restraint, but we believe that such external controls can never be so subtle or so effective as internal controls. It would therefore be sensible to promote the social conditions in which public opinion and full public participation in decision-making become as far as possible the means whereby communities are ordered. The larger a community the less likely this can be: in a heterogeneous, centralized society such as ours, the restraints of the stable society if they were to be effective would appear as so much outside coercion; but in communities small enough for the general will to be worked out and expressed by individuals confident of themselves and their fellows as individuals, 'us and them' situations are less likely to occur – people having learned the limits of a stable society would be free to order their own lives within them as they wished, and would therefore accept the restraints of the stable society as necessary and desirable and not as some arbitrary restriction imposed by a remote and unsympathetic government.

2. As agriculture depends more and more on integrated control and becomes more diversified, there will no longer be any scope for prairie-type crop-growing or factory-type livestock-

rearing. Small farms run by teams with specialized knowledge of ecology, entomology, botany, etc. will then be the rule, and indeed individual small-holdings could become extremely productive suppliers of eggs, fruit and vegetables to neighbour-hoods. Thus a much more diversified urban–rural mix will be not only possible but, because of the need to reduce the trans-portation costs of returning domestic sewage to the land, desir-able. In industry, as with agriculture, it will be important to maintain a vigorous feedback between supply and demand in order to avoid waste, overproduction, or production of goods which the community does not really want, thereby eliminating the needless expense of time, energy and money in attempts to persuade it that it does. If an industry is an integral part of a community, it is much more likely to encourage product innova-tion because people clearly want qualitative improvements in a given field, rather than because expansion is necessary for that industry's survival or because there is otherwise insufficient work for its research and development section. Today, men, women and children are merely consumer markets, and indus-tries as they centralize become national rather than local and supranational rather than national, so that while entire com-munities may come to depend on them for the jobs they supply, they are in no sense integral parts of those communities. To a considerable extent the 'jobs or beauty' dichotomy has been made possible because of this deficiency. Yet plainly people want jobs *and* beauty, they should not in a just and humane society be forced to choose between the two, and in a decentral-ized society of small communities where industries are small enough to be responsive to each community's needs there will be no reason for them to do so.

3. The small community not only is the organizational struc-ture in which internal or systemic controls are most likely to operate effectively, but its dynamic is an essential source of stimulation and pleasure for the individual. Indeed it is probable that only in the small community can a man or woman be an individual. In today's large agglomerations he is merely an isolate – and it is significant that the decreasing autonomy of communities and local regions and the increasing centralization

of decision-making and authority in the cumbersome bureau-cracies of the state, have been accompanied by the rise of self-conscious individualism, an individualism which feels threatened unless it is harped upon. Perhaps the two are mutually dependent. It is no less significant that this self-conscious individualism tends to be expressed in ways which cut off one individual from another – for example the accumulation of material goods like the motor-car, the television set, and so on, all of which tend to insulate one from another, rather than bring them together. In the small, self-regulating communities observed by anthropologists, there is by contrast no assertion of individualism, and certain individual aspirations may have to be repressed or modified for the benefit of the community – yet no man controls another and each has very great freedom of action, much greater than we have today. At the same time they enjoy the rewards of the small community, of knowing and being known, of an intensity of relationships with a few, rather than urban man's variety of innumerable, superficial relationships. Such rewards should provide ample compensation for the decreasing emphasis on consumption, which will be the inevitable result of the premium on durability which we have suggested should be established so that resources may be conserved and pollution minimized. This premium, while not diminishing our real standard of living, will greatly reduce the turnover of material goods. They will thus be more expensive, although once paid for they should not need replacing except after long periods. Their rapid accumulation will no longer be a realizable or indeed socially acceptable goal, and alternative satisfactions will have to be sought. We believe a major potential source of these satisfactions to be the rich and variegated interchanges and responsibilities of community life, and that these are possible only when such communities are on a human scale.

4. The fourth reason for decentralization is that to deploy a population in small towns and villages is to reduce to the minimum its impact on the environment. This is because the actual urban superstructure required per inhabitant goes up radically as the size of the town increases beyond a certain point. For example, the *per capita* cost of high rise flats is much greater

than that of ordinary houses; and the cost of roads and other transportation routes increases with the number of commuters carried. Similarly, the *per capita* expenditure on other facilities such as those for distributing food and removing wastes is much higher in cities than in small towns and villages. Thus, if everybody lived in villages the need for sewage treatment plants would be somewhat reduced, while in an entirely urban society they are essential, and the cost of treatment is high. Broadly speaking, it is only by decentralization that we can increase self-sufficiency – and self-sufficiency is vital if we are to minimize the burden of social systems on the ecosystems that support them.

Although we believe that the small community should be the basic unit of society and that each community should be as self-sufficient and self-regulating as possible, we would like to stress that we are not proposing that they be inward-looking, self-obsessed or in any way closed to the rest of the world. Basic precepts of ecology, such as the interrelatedness of all things and the far-reaching effects of ecological processes and their disruption, should influence community decision-making, and therefore there must be an efficient and sensitive communications network between all communities. There must be procedures whereby community actions that affect regions can be discussed at regional level and regional actions with extra-regional effects can be discussed at global level. We have no hard and fast views on the size of the proposed communities, but for the moment we suggest neighbourhoods of 500, represented in communities of 5,000, in regions of 500,000, represented nationally, which in turn as today should be represented globally. We emphasize that our goal should be to create *community feeling* and *global awareness*, rather than that dangerous and sterile compromise which is nationalism.

In many of the developed countries where community feeling has been greatly eroded and has given way to heterogeneous congeries of strangers, the task of re-creating communities will be immensely difficult. In many of the undeveloped countries, however, although it will not be easy, because the process of community collapse and flight to the city has begun only recently there is a real chance that it can be halted by such means as the abandonment of large-scale industrial projects for the development of intermediate technologies at village level; and the provision of agro-ecological training teams so that communities can be taught to manage the land together, rather than encourage farmers to turn to expensive and dangerous procedures like the heavy use of pesticides and fertilizers, which tend to reduce the number of people needed on the land.

At home, industry will play a leading role in the programme to decentralize our economy and society. The discussion of taxes, anti-disamenity legislation, and enforceable targets for air, land and water quality in the section on stock economics might lead some to believe that we are willing to bring about the collapse of industry, widespread unemployment, and the loss of our export markets. It is therefore worth emphasizing that we wish strongly to avoid all three, and we do not see that they are necessary or inevitable consequences of our proposals. It is obvious that for as long as we depend on imports for a significant proportion of our food, so we must export. And since we are likely to require food-imports for the next 150 years, we are left with the question of whether it is possible to develop community industries, dedicated to the principles of maximal use/recycling of materials and durability of goods, and at the same time to earn an adequate revenue from exports.

We believe that the answer is yes, if the change-over is conducted in two stages. The first stage is to alter the direction of growth so that it becomes more compatible with the aims of a stable society. We have already mentioned that the recycling industry must be encouraged to expand, and it is obvious that willy-nilly it will do so as over the years taxes and quality targets become more stringent. To give a clearer idea of how the direction can be altered we will consider briefly the question of transport.

There are more than 12 million cars in Britain today, and according to the Automobile Association this figure will rise to 21 million by 1981. About half the households in Britain own a car today, and presumably the car population is expected to rise in response to a rise in this proportion, though presumably, too, more households will own more than one car. At all events we have sufficient experience of traffic congestion in our towns and cities and the rape of countryside and community by ring-roads and motorways to realize that the motor-car is by no means the best way of democratizing mobility. Indeed, if every household had a car, we would be faced with the choice of leaving towns and country worth driving to and thereby imposing immobility on the motorist, or of providing him with the vast expanses of concrete which are becoming increasingly necessary to avoid congestion at the expense of the areas they sterilize and blight.

No one can contemplate with equanimity the doubling of roads within this decade necessary to maintain the *status quo*, and we must therefore seek sensible transportation alternatives. It is clear that broadly speaking the only alternative is public transport – a mix of rapid mass-transit by road and rail. Rail especially should never have been allowed to run down to the extent that it has. The power requirements for transporting freight by road are five to six times greater than by rail and the pollution is correspondingly higher. The energy outlay for the cement and steel required to build a motorway is three to four times greater than that required to build a railway, and the land area necessary for the former is estimated to be four times more than for the latter. Public transport whether by road or rail is much more efficient in terms of *per capita* use of materials and energy than any private alternative. It can also be as flexible, provided it is encouraged at the expense of private transport.

This is the key to the provision of a sound transportation system. First the vicious spiral of congestion slowing buses, losing passengers, raising fares, losing more passengers, using more cars, creating more congestion, etc. must be broken. A commitment to build no more roads and to use the capital released to subsidize public transport would be an excellent way of doing this. The men who would normally live by road-building could be diverted to clearing derelict land and restoring railways and canals as part of a general programme of renewal. From there, the progressive imposition of restrictions on private transport and the stimulation of public transport so that it could provide a fast, efficient and flexible alternative would be a matter of course. Within the motor industry, the decline in production of conventional private vehicles would be compensated for by the increased production of alternative mass-transit systems. There would also be a switch of capital and manpower to the redevelopment of railway systems. In the long term, however, decentralization will bring a diminished demand for mobility itself. As Stephen Boyden has pointed out (7), people use their cars for four main reasons: to go to work, to go to the countryside, to visit friends and relations, and to show off. In the stable society, however, each community will provide its own jobs, there will be countryside around it, most friends and relations will be within it, and there will be much more reliable and satisfying ways of showing off.

This brings us to the second stage of the change-over, in which industry turns to the invention, production, and installation of technologies that are materials and energy conservative, that are flexible, non-polluting and durable, employment-intensive and favouring craftsmanship. Progress as we conceive of it today consists in increasing an already arbitrarily high ratio of capital to job availability; but if instead this ratio were to be reduced, then our manpower requirement would go up, while at the same time the pollution which is the inevitable by-product of capital growth would be cut down. The switch in emphasis from quantity to quality will not only stimulate demand for manpower, it will also stabilize it *and* give much greater satisfaction to the men themselves. Instead of men being used as insensate units to produce increasing quantities of components, they should be trained and given the opportunity to improve the quality of their work. The keynotes of the manufacturing sector should come to be durability and craftsmanship – and such a premium on quality should assure us an export revenue large enough for us to continue buying food from abroad, while providing our manpower with more enjoyable occupations. In the case of industries like the aircraft industry, which would naturally have a greatly reduced role in the stable society, their engineering expertise could be turned to the development of such things as total energy systems – designed to provide the requirements of a decentralized society with the minimum of environmental disruption.

Industry can completely fulfil its new role only in close harmony with particular communities, so that the unreal distinction between men as employees and men as neighbours can be abandoned, and jobs then given on the basis that work must be provided by the community for the sake of that community's stability and not because one group wishes to profit from another group's labour or capital as the case may be. As industry decentralizes so will the rest of society. The creation of communities will come from the combination of industrial change and a conscious drive to restructure society.

The principal components of this drive are likely to be the redistribution of government and the gradual inculcation of a sense of community and the other values of a stable society. Over a stated period of time, local government should be strengthened and as many functions as possible of central government should be transferred to it. The redistribution of government should proceed on the principle that issues which affect only neighbourhoods should be decided by the neighbourhood alone, those which affect only communities by the community alone, those which affect only regions by the region alone, and so on. As regions, communities and neighbourhoods come increasingly to run their own affairs, so the development of a sense of community will proceed more easily, though we do not pretend that it will be without its problems.

Those regions which still have or are close to having a good urban–rural mix will be able to effect a relatively smooth transfer, but highly urbanized areas like London, the Lancashire conurbation, and South Wales will find it much more difficult to re-create communities. Nevertheless, even in London the structural remains of past communities (like the villages of Putney, Highgate, Hackney, Islington, etc.) will provide the physical nuclei of future communities – the means of orienting themselves so that they can cut themselves away from those deserts of commerce and packaged pleasure (of which the most prominent example is the Oxford Street, Regent Street, Piccadilly complex) on which so much of London's life is currently focused.

It is self-evident that no amount of legislative, administrative or industrial change will create stable communities if the individuals who are meant to comprise them are not fitted for them. As soon as the best means of inculcating the values of the stable society have been agreed upon, they should be incorporated into our educational systems. Indeed, it may not be until the generation of 40–50 year olds have been educated in these values (so that as far as possible everybody up to the age of 50 understands them) that stable communities will achieve sufficient acceptance for them to be permanently useful.

ORCHESTRATION

A cardinal assumption of this strategy is that it will not succeed without the most careful synchronization and integration. We cannot say of a particular section of these proposals that it alone is acceptable, and therefore we will go ahead with it immediately but consider the rest later on! This section, therefore, is devoted to a schematic, annotated outline of how change might be orchestrated (see Figure 3). It is necessarily unsophisticated and oversimplified, but we hope it will give some idea of how change in one quarter will aid change in the others.

Variables included in schematic outline:

(a) establishment of national population service;

(b) introduction of raw materials, amortization and power taxes; anti-disamenity legislation; air, land and water quality targets; recycling grants; revised social accounting systems;

(c) developed countries end commitment to persistent pesticides and subsidize similar move by undeveloped countries;

(d) end of subsidies on inorganic fertilizers;

(e) grants for use of organics and introduction of diversity;

(f) emergency food programme for undeveloped countries;

(g) progressive substitution of non-persistent for persistent pesticides;

(h) integrated control research programme;

(i) integrated control training programme;

(j) substitution of integrated control for chemical control;

(k) progressive introduction of diversified farming practices;

(l) end of road building;

(m) clearance of derelict land and beginning of renewal programme;

(n) restrictions on private transport and subsidies for public transport;

(o) development of rapid mass-transit;

(p) research into materials substitution;

(q) development of alternative technologies;

(r) decentralization of industry: part one (redirection);

(s) decentralization of industry: part two (development of community types);

(t) redistribution of government;

(u) education research;

(v) teacher training;

(w) education;

(x) experimental community;

(y) domestic sewage to land;

(z) target date for basic establishment of network of self-sufficient, self-regulating communities.

Notes:

(1) should be operating fully by 1980; review in 1995 – if replacement-size families improbable by 2000, bring in socio-economic restraints; UK population should begin to decline slowly from A.D. 2015–20 onwards; world population from 2100; little significant feedback expected in UK until about 2030.

(2) is progressive; ironing-out run to eliminate inconsistencies up to 1980; thereafter revise and tighten every five years; increasingly significant feedback from 1980 onwards, stimulating materials-energy conservation, employment-intensive industry, decentralization, and progress in direction of (p), (q), (r) and (s).

(3) Limited substitution of integrated control can begin quite soon, but large-scale substitution will depend on (h) integrated control research programme; naturally (h), (i) and (j) will run in parallel and are therefore represented as one; (g) will also continue for some time.

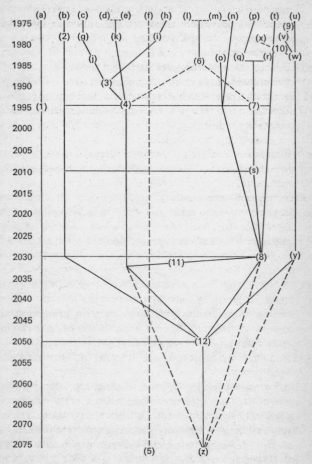

Figure 3 Schematic outline of change.
(See text for key.)

(4) Diversified farming practices (k) and integrated control (j) will link up and form an agriculture best suited for small, reasonably self-sufficient communities, so stimulating their development: significant feedback, therefore, will occur from this point.

(5) is likely to be necessary at least until 2100.

(6) Labour released from road-building can go to (m) clearance of derelict land, which should be completed by 1985; thereafter there may be other renewal programmes such as canal restoration, while agriculture will increasingly require more manpower.

(7) Development of alternative technologies (q) and redirecting of industry (r) will proceed in harness; progressively significant feedback between (b) and (t).

(8) Target date for maximum redistribution of government 2030 to coincide with 45 years operation of (w); see Note (9).

(9) Five years only allowed for preliminary organization and research, since it can proceed in harness with teacher training (v) and also with the education programme itself (w).

(10) An experimental community of 500 upwards could be set up to clarify problems; feedback to (u).

(11) As soon as communities are small enough, domestic sewage can be returned to the land; there should be the firm beginnings of a good urban–rural mix by then.

(12) By this time there should be sufficient diversity of agriculture, decentralization of industry and redistribution of government, together with a large proportion of people whose education is designed for life in the stable society, for the establishment of self-sufficient, self-regulating communities to be well advanced. At this point taxation, grants, incentives, etc., could be taken over by the communities themselves. A further generation is allowed until target date, however.

3 The Goal

There is every reason to suppose that the stable society would provide us with satisfactions that would more than compensate for those which, with the passing of the industrial state, it will become increasingly necessary to forgo.

We have seen that man in our present society has been deprived of a satisfactory social environment. A society made up of decentralized, self-sufficient communities, in which people work near their homes, have the responsibility of governing themselves, of running their schools, hospitals, and welfare services, in fact of constituting real communities, should, we feel, be a much happier place. Its members, in these conditions, would be likely to develop an identity of their own, which many of us have lost in the mass society we live in. They would tend, once more, to find an aim in life, develop a set of values, and take pride in their achievements as well as in those of their community.

It is the absence of just these things that is rendering our mass society ever less tolerable to us and in particular to our youth, and to which can be attributed the present rise in drug-addiction, alcoholism and delinquency, all of which are symptomatic of a social disease in which a society fails to furnish its members with their basic psychological requirements.

More than a hundred years ago, John Stuart Mill realized that industrial society, by its very nature, could not last for long and that the stable society that must replace it would be a far better place. He wrote (1):

I cannot . . . regard the stationary state of capital and wealth with the unaffected aversion so generally manifested towards it by political economists of the old school. I am inclined to believe that it would be, on the whole, a very considerable improvement on our present condition. I confess I am not charmed with the ideal of life held out by those who think that the normal state of human beings is that of struggling to get on; that the trampling, crushing, elbowing, and treading on each other's heels which forms the existing type of social life, are the most desirable lot of human kind... The northern and middle states of America are a specimen of this stage of civilisation in very favourable circumstances; and all that these advantages seem to have yet done for them ... is that the life of the whole of one sex is devoted to dollar-hunting, and of the other to breeding dollar-hunters.

I know not why it should be a matter of congratulation that persons who are already richer than anyone needs to be should have doubled their means of consuming things which give little or no pleasure except as representative of wealth ... It is only in the backward countries of the world that increased production is still an important object; in those most advanced, what is economically needed is a better distribution, of which one indispensable means is a stricter restraint on population ... The density of population necessary to enable mankind to obtain, in the greatest degree, all the advantages both of cooperation and of social intercourse, has, in all the most populous countries, been attained ... It is not good for a man to be kept perforce at all times in the presence of his species ... Nor is there much satisfaction in contemplating a world with nothing left to the spontaneous activity of nature ... If the earth must lose that great portion of its pleasantness which it owes to things that the unlimited increase of wealth and population would extirpate from it, for the mere purpose of enabling it to support a larger population, I sincerely hope, for the sake of posterity, that they will be content to be stationary, long before necessity compels them to it.

It is scarcely necessary to remark that a stationary condition of capital and population implies no stationary state of human improvement. There would be as much scope as ever for all kinds of mental culture, and moral and social progress; as much room for improving the Art of Living and much more likelihood of it being improved, when minds ceased to be engrossed by the art of getting on.

THE IMPORTANCE OF A VARIED ENVIRONMENT

In our industrial society, the only things that tend to get done are those that are particularly conducive to economic growth, those in fact that, in terms of our present accounting system, are judged most efficient! This appears to be almost the sole consideration determining the nature of the crops we sow, the style of our houses, and the shape of our cities. The result, among other things, is the dreariest possible uniformity.

In a stable society, on the other hand, there would be nothing to prevent many other considerations from determining what we cultivate or build. Diversity would thus tend to replace uniformity, a trend that would be accentuated by the diverging cultural patterns of our decentralized communities. As René Dubos has pointed out (2):

In his recent book, *The Myth of the Machine*, Lewis Mumford states that 'If man had originally inhabited a world as blankly uniform as a "high-rise" housing development, as featureless as a parking lot, as destitute of life as an automated factory, it is doubtful that he would have had a sufficiently varied experience to retain images, mould language, or acquire ideas.' To this statement, Mr Mumford would probably be willing to add that, irrespective of genetic constitution, most young people raised in a featureless environment and limited to a narrow range of life experiences will be crippled intellectually and emotionally.

We must shun uniformity of surroundings as much as absolute conformity of behaviour, and make instead a deliberate effort to create as many diversified environments as possible. This may result in some loss of efficiency, but the more important goal is to provide the many kinds of soil that will permit the germination of the seeds now dormant in man's nature. In so far as possible, the duplication of uniformity must yield to the organisation of diversity. Richness and variety of the physical and social environment constitute crucial aspects of functionalism, whether in the planning of cities, the design of dwellings, or the management of life.

REAL COSTS

We might regard with apprehension a situation in which we shall have to make do without many of the devices such as motorcars, and various domestic appliances which, to an ever greater extent, are shaping our everyday lives.

These devices may indeed provide us with much leisure and satisfaction, but few have considered at what cost. For instance, how many of us take into account the dull and tedious work that has to be done to manufacture them, or for that matter to earn the money required for their acquisition? It has been calculated (3) that the energy used by the machines that provide the average American housewife with her high standard of living is the equivalent of that provided by five hundred slaves.

. In this respect, it is difficult to avoid drawing a comparison between ourselves and the Spartans, who in order to avoid the toil involved in tilling the fields and building and maintaining their homes employed a veritable army of helots. The Spartan's life, as everybody knows, was a misery. From early childhood, boys were made to live in barracks, were fed the most frugal and austere diet and spent most of their adult life in military training so as to be able to keep down a vast subject population, always ready to seize an opportunity to rise up against its masters. It never occurred to them that they would have been far better off without their slaves, fulfilling themselves the far less exacting task of tilling their own fields and building and maintaining their own homes.

In fact 'economic cost', as we have seen, simply does not correspond to 'real cost'. Within a stable society this gap must be bridged as much as possible.

This means that we should be encouraged to buy things whose production involves the minimum environmental disruption and which will not give rise to all sorts of unexpected costs that would outweigh the benefits that their possession might provide.

REAL VALUE

It is also true, as we have seen, that 'economic value' as at present calculated does not correspond to real value any more than 'economic cost' corresponds to real cost.

Our standard of living is calculated in terms of the market prices of the goods that it includes. These do not distinguish between, on the one hand, the gadgets that we do not really need and such essentials as unpolluted water, air and food on which our health must depend. In fact it tends to place greater value on the former, as we usually take the latter for granted.

It is in terms of these market prices that the GNP is calculated, and, as we have seen, this provides the most misleading indication of our well-being. Edward Mishan (4) points out that

... an increase in the numbers killed on the roads, an increase in the numbers dying from cancer, coronaries or nervous diseases, provides extra business for physicians and undertakers, and can contribute to raising GNP. A forest destroyed to produce the hundreds of tons of paper necessary for the American Sunday editions is a component of GNP. The spreading of concrete over acres of once beautiful country-side adds to the value of GNP ... and so one could go on.

In the same way, many of the machines whose possession is said to increase our standard of living are simply necessary to replace natural benefits of which we have been deprived by demographic and economic growth. We have pointed out how true this is of the ubiquitous motor-car. Also, many labour-saving devices are now necessary because with the disintegration of the extended family there is no one about to do the household chores. The fact that both husband and wife must, in many cases, go out to work to earn the money to buy the machines required to do these chores can serve only to render such devices that much more necessary.

In a stable society, everything would be done to reduce the discrepancy between economic value and real value, and if we could repair some of the damage we have done to our physical and social environment, and live a more natural life, there would be less need for the consumer products that we spend so much money on. Instead we could spend it on things that truly enrich and embellish our lives.

In manufacturing processes, the accent would be on quality rather than quantity, which means that skill and craftsmanship, which we have for so long systematically discouraged, would once more play a part in our lives. For example, the art of cooking would come back into its own, no longer regarded as a form of drudgery, but correctly valued as an art worthy of occupying our time, energy and imagination. Food would become more varied and interesting, and its consumption would become more of a ritual and less a utilitarian function.

The arts would flourish: literature, music, painting, sculpture and architecture would play an ever greater part in our lives, while achievements in these fields would earn both money and prestige.

A society devoted to achievements of this sort would be an infinitely more agreeable place than is our present one, geared as it is to the mass production of shoddy utilitarian consumer goods in ever greater quantities. Surprising as it may seem to one reared on today's economic doctrines, it would also be the one most likely to satisfy our basic biological requirements for food, air and water, and even more surprisingly, provide us with the jobs that in our unstable industrial society are constantly being menaced.

Indeed, as we have seen, the principal limitation to the availability of jobs today is the inordinately high capital outlay required to finance each worker. This limitation is withdrawn as soon as we accept that, within the framework of an overall reorganization of our society, it would be possible for capital outlay to be reduced without reducing our real standard of living.

One of the Bishop of Kingston's ten commandments (5) is: 'You shall not take the name of the Lord thy God in vain by calling on his name but ignoring his natural law.' In other words, there must be a fusion between our religion and the rest of our culture, since there is no valid distinction between the laws of God and Nature, and Man must live by them no less than any other creature. Such a belief must be central to the philosophy of the stable society, and must permeate all our thinking. Indeed it is the only one which is properly scientific, and science must address itself much more vigorously to the problems of co-operating with the rest of Nature, rather than seeking to control it.

This does not mean that science must in any way be discouraged. On the contrary, within a stable society there would be considerable scope for the energies and talents of scientist and technologist.

Basic scientific research, plus a good deal of multidisciplinary synthesis, would be required to understand the complex mechanisms of our ecosphere with which we must learn to co-operate.

There would be a great demand for scientists and technologists capable of devising the technological infrastructure of a decentralized society. Indeed, with the application of a new set of criteria for judging the economic viability of technological devices, there must open a whole new field of research and development.

The recycling industry which must expand very considerably would offer innumerable opportunities, while in agriculture there would be an even greater demand for ecologists, botanists, entomologists, mycologists etc., who would be called upon to devise ever subtler methods for ensuring the fertility of the soil and for controlling 'pest' populations.

Thus in many ways, the stable society, with its diversity of physical and social environments, would provide considerable scope for human skill and ingenuity.

Indeed, if we are capable of ensuring a relatively smooth transition to it, we can be optimistic about providing our children with a way of life psychologically, intellectually and aesthetically more satisfying than the present one. And we can be confident that it will be sustainable as ours cannot be, so that the legacy of despair we are about to leave them may at the last minute be changed to one of hope.

Appendix A:
Ecosystems and their disruption

It is necessary to survey the essential features of the environment in order to understand how it is being affected by man's activities.

We can define the environment as a system which includes all living things and the air, water and soil which is their habitat. This system is often referred to as the ecosphere. To describe it as a system is to accentuate its unity; a system being something made up of interrelated parts in dynamic interaction with each other, and capable, for certain purposes, of co-operating in a common behavioural programme.

Such a programme must be regarded as goal-directed, and its goal the maintenance of stability. This appears to be the basic goal of all the self-regulating behavioural processes that make up the ecosphere.

Stability is best defined as a system's ability to maintain its basic features – in other words to survive in the face of environmental change. This means that, in a stable system, change will be minimized and will occur only as is necessary to ensure adaptation to a changing environment. In other words, as stability increases so the frequency of random changes will be correspondingly reduced.

It is easy to see how the ecosphere during the last few thousand million years of evolution has slowly become more stable.

Whereas the deserts, which once covered our planet, reflected the environmental pressures to which they were subjected, the forests that developed to replace them have a capacity to maintain a relatively stable situation in the face of internal and external change. For instance, they ensure an optimum balance between the oxygen and carbon dioxide contents of the air by emitting one and absorbing the other. They provide good

conditions for the run-off to rivers to be regulated. They period-
ically shed their leaves, which build up humus and hence ensure
the continued fertility of the soil. They provide a relatively
constant ambient temperature to the wild animals that live
within their shade, who, as they evolve, also develop stabilizing
mechanisms ensuring the stability of what is sometimes called
their 'internal environment'; the constant body temperature
of warm-blooded mammals being an obvious example.

Perhaps the most important feature of the ecosphere is its
degree of organization. It is made up of countless ecosystems,
themselves organized into smaller ones, which are further
organized into still smaller ones. Each of these is made up of
populations of different species in close interaction with each
other, some of which are usually organized into communities and
families – further organized into cells, molecules and atoms, etc.

The opposite of organization is randomness, or what is often
referred to as entropy. In fact it can be said that the ecosphere
differs from the surface of the moon, and probably from that of
all the other planets in our solar system, in that randomness, or
entropy, has been progressively reduced and organization, or
negative entropy, has been correspondingly increased. According
to the second law of thermodynamics, there is a tendency in all
systems towards increasing randomness, or entropy. This must
be so, since to move in this direction is to take the line of least
resistance, and also because whenever energy is converted (and
this must occur during all behavioural processes), waste, or
random parts, must be generated – from oxidation and friction
if from nothing else.

The ecosphere has succeeded in counteracting this tendency
by virtue of several unique features and because it is an open
system from the point of view of energy, being continually
bombarded with solar radiation.

This radiation is used by green plants during photosynthesis
to organize nutrients in the soil into complex plant tissue, which
are then eaten by herbivores, and hence reorganized into still
more complex animal tissue.

In such processes waste or random parts must be generated.
However, so long as the corresponding reduction in organization

is less than the increase in organization achieved during the process, then entropy will have been reduced. Such increases will be limited by all sorts of factors, including the availability of energy and materials, the environment's capacity to absorb waste and the organizational capacity of the system. Waste must therefore be kept down to a minimum. This can be done only by recycling it so as to ensure that the waste generated by one process serves as the materials for the next. This is essential for another reason:

Whereas the ecosphere is an open system as regards energy, it is a closed one as regards materials, which is another reason why all materials must be recycled, and why the waste products of one process must serve as materials for the next.

Also some of the more highly organized materials required for sophisticated processes have taken hundreds of millions of years to develop in the case of fossil fuels, for instance, and thousands of millions of years in the case of the herbivorous animals required as food by carnivores. It is thus clear that to avoid increasing entropy they cannot be used up faster than they are produced. Hence the essential cyclic nature of all ecological processes and the absolute necessity for recycling everything.

It is possible to trace just how all the resources, such as carbon, nitrogen, phosphorus, water, etc., made use of in behavioural processes are recycled. The food cycle is particularly illustrative. Take the case of a marine ecosystem: fish excrete organic waste which is converted by bacteria to inorganic products. These provide nutrients, permitting the growth of algae, which are eaten by fish; and the cycle is complete. In this way the wastes are eliminated, the water kept pure, and, at the same time, the materials for the next stage of the process are made available.

One of the most important features of life processes is that they are automatic or self-regulating. Self-regulation can be ensured in only one way: data must be detected by the system, transduced into the appropriate informational medium, and organized so as to constitute a model or 'template' of its relationship with its environment. Whenever this relationship is modified in such a way that it deviates from the optimum, the model is correspondingly affected, and it can be used to guide the appropriate course of action, and monitor each new move, until a new position of equilibrium has been reached. This basic cybernetic model explains how all systems, regardless of their level of complexity, adapt to their respective environments. The fact that all the parts of the ecosphere are linked to each other in this way ensures that a general readjustment of the most subtle nature can occur to restore its basic structure after any disturbance.

To suppose that we can ensure the functioning of the ecosphere ourselves with the sole aid of technological devices, thereby dispensing with the elaborate set of self-regulating mechanisms that has taken thousands of millions of years to evolve, is an absurd piece of anthropocentric presumption that belongs to the realm of pure fantasy. It may be possible to replace certain natural controls locally and for a short while without any serious cataclysm occurring, but if we push things too far, if for instance the insecticides we use to replace the self-regulating controls that normally ensure the stability of insect populations were to destroy nitrogen-fixing bacteria or pollinating insects, all the money and all the technology in the world would not suffice to replace them and thereby to prevent life processes from grinding to a halt. Yet this substitution is implicit in the aim of industrial society.

As this aim is progressively realized, and as we become more and more dependent on technological devices, i.e. external controls, so must there be a corresponding increase in the instability of our social system and hence in our vulnerability to change. Imagine what it will be like when water supplies have been exhausted and we are dependent upon desalination plants for our drinking water, when traditional methods of agriculture have totally given way to ever more ingenious forms of factory farming, and when the natural mechanisms providing us with the air we breathe have been so completely disrupted that vast installations are needed to pump oxygen into the atmosphere and filter out the noxious gases emitted by our industrial installations.

Clearly, under such conditions the slightest technical hitch or industrial dispute, or shortage of some key resource, might be sufficient to deprive us of such basic necessities of life as water, food and air – and bring life to a halt.

If man wishes to survive, to ensure the proper functioning of the self-regulating mechanisms of the ecosphere must be his most basic endeavour. For this to be possible however the latter's essential structure must be respected. Deviations may be possible, but only within certain limits.

One way of exceeding these limits is to supply the system with more waste than can be used to provide the materials for other processes. In such conditions the system is said to be 'overloaded'; the self-regulating mechanisms can no longer function and the waste simply accumulates. In other words entropy, or randomness, has increased and the surface of the earth resembles that much more that of the moon. Thus, to return to our marine ecosystem, if the cycle is overloaded with too much sewage, detergents or artificial fertilizers which are nutrients to aquatic plant life, the amount of oxygen required to ensure the decomposition of these substances by the appropriate bacteria may be so high that other organisms will be deprived of an adequate supply. If this goes on long enough the oxygen level will be reduced to zero. Without oxygen, the bacteria will die and a crucial phase in the cycle will have been interrupted, thereby bringing it rapidly to a halt. As a result, what was once an elaborate ecosystem, supporting countless forms of life in close interaction with each other, now becomes a random arrangement of waste matter.

Needless to say the cycle will also come to a halt if, on the contrary, there were a shortage of nutrients. In such conditions the algae could not survive, and the fish population, deprived of its sustenance, would rapidly die off.

This illustrates an essential principle of organization: there must be an optimum value to every variable in terms of which the system is described. When each variable has its correct value, then the system described can be regarded as having its correct structure. This means that there is no value that can be increased or reduced indefinitely without bringing about the system's eventual breakdown.

To cherish the illusion that the population and affluence of human social systems are exceptions to this law is, as we shall see, to court the gravest possible calamities.

In order to maintain the system's structure, the actions of the self-regulating sub-systems seek to establish a stable relationship not only with another sub-system, but with their environment as a whole. In other words, they do not aim at satisfying a specific requirement, but at achieving a compromise between a whole set of often competing requirements: that which best satisfies the requirements of the environment as a whole.

Technological devices, of course, do precisely the opposite. They are geared to the achievement of specific short-term targets regardless of environmental consequences. Since many requirements must be satisfied to maintain stability, such devices by their very nature must cause environmental problems, and, as a result, they must inevitably tend towards achieving equilibrium positions which display lower rather than higher stability. This means that the probability that disequilibria will occur, and their degrees of seriousness, are both likely to increase, as must the rate at which new devices will be required as well as the effectiveness required of them. In other words, the role played by technology must increase by positive feedback and our society must become ever more addicted to it.

In these circumstances, unless technological innovation can proceed indefinitely at an exponential rate, then it is only a question of time before a disequilibrium occurs for which there is no technological solution, which must spell the complete breakdown of the system.

Industrial society, when it reaches a certain stage of development, begins to affect its environment in yet another manner; it devises, and becomes correspondingly dependent upon, synthetic products of different sorts to replace ever-scarcer natural products. Thus plastics are developed to replace wood products; detergents to replace soaps made from natural fats; synthetic fibres to replace natural fibres; chemical fertilizers to replace organic manure. At the same time, nuclear energy slowly replaces that previously derived from fossil fuels. It is probable that our ecosphere does not produce a single molecule for which there is not an enzyme capable of breaking it down, in order to perpetuate the essential cycle of life, growth, death and decay. This is not so with synthetic products. They cannot normally be broken down in this way – save in some cases by human manipulation, which is practicable on a small scale only and in specific conditions. It is thus no longer a question of overloading a system. Even the slightest amount of these products, when introduced into our ecosphere, constitutes pollution, while, since by their very nature they must continue accumulating, to produce them methodically is to ensure the systematic replacement of the ecosphere with extraneous waste matter.

What is worse, many of these substances find their way into life processes with which they can seriously interfere. Thus strontium-90 gets into the bones of growing children and can give rise to bone cancer; iodine-131 accumulates in the thyroid gland and can give rise to cancer of the thyroid; DDT accumulates in the fatty matter and in the liver and may cause cancer and other liver diseases; plastics and many other pollutants also accumulate in the liver and kidneys, and so on.

It is not surprising that as industrialization proceeds so there is a very rapid increase in the so-called degenerative diseases. Carcinogenic agents also tend to be mutagens, and their proliferation must mean a gradual reduction in the adaptiveness of our species, a process that clearly cannot go on indefinitely (2).

There is another way in which we are degrading the ecosphere. One of its most important features is its complexity. The greater number of different plant and animal species that make up an ecosystem, the more likely it is to be stable. This is so because, as Elton points out, in such a system every ecological niche is filled. That is to say, every possible differentiated function for which there is a demand within the system is in fact fufilled by a species that is specialized in fulfilling it. In this way it is extremely difficult for an ecological invasion to occur, i.e. for a species foreign to the system to enter and establish itself, or, worse still, to proliferate and destroy the system's basic structure.

It also means that no species forming part of the system is likely to be able to expand beyond its optimum size. The availability and size of an ecological niche undoubtedly constitutes an effective population control. Thus the diet of a specialized member of a highly differentiated ecosystem will itself be of a specialized nature, which means that if the population of a particular species were to increase, or, alternatively, to decrease, the food supply of the other species would not be affected. The opposite would be the case with species that normally form part of a simple ecosystem. Thus goats are adapted to live in mountain areas, where ecological complexity is low, and in order to survive they have to be able to eat almost anything. The result is when they are brought down to the plains they make short shrift of its vegetation, and their proliferation compromises the food supply of many other species.

As industrial man destroys the last wildernesses, as herds of domesticated animals replace interrelated animal species, and vast expanses of crop monoculture supplant complex plant ecosystems, so complexity and hence stability are correspondingly reduced.

Industrial man is also reducing complexity in other ways. For instance, economic pressures force farmers to reduce the number of different strains of crops under cultivation. Only those that present short-term economic advantages tend to survive. This process has been accentuated with the so-called 'green revolution'. Special high-yield strains of rice and wheat that respond particularly well to artificial fertilizers have been developed and introduced on a large scale in many parts of the third world. In these areas many other strains have been abandoned. In this way we are reducing complexity, in some cases irreversibly, and if anything should happen to the surviving strains essential crops like wheat and rice could well be jeopardized.

We are reducing complexity in still another way. The greater the number of trophic levels (in other words the greater the length of food chains), the more stable is an ecosystem likely to be. Thus the simplest marine ecosystem would consist of phytoplankton, capable of harnessing the sun's energy, and microorganisms capable of decomposing them. By introducing zooplankton into the system, another link has been introduced into the food chain. These, by preying on the phytoplankton, keep down their numbers and weed out the weak and unadaptive. In this way, they exert both quantitative and qualitative controls, and exert an important stabilizing influence. If fish are then introduced to feed on the zooplankton, the system becomes correspondingly more stable.

Needless to say, man's activities are everywhere leading to a reduction in the length of food chains. The larger terrestrial predators have been virtually eliminated in industrial countries, and this process is now taking place in the seas. Man, by refusing to tolerate competitors for his food supply, is ultimately jeopardizing the stability of this food supply and, hence, its very availability.

Also, as SCEP points out, environmental stress appears to affect predators more radically than herbivores. In aquatic systems the top-level predators, which eat other predators, are the most sensitive of all. This appears to be the case with such disruptive situations as oxygen deficiency, thermal stress, and the introduction of toxic materials such as pesticides and fertilizers.

The effect must be to reduce the number of trophic levels in any ecosystem, thereby increasing its instability. SCEP cites several examples:

Overenrichment by sewage waste and fertiliser runoff of fresh-waters, or pollution with industrial wastes, leads to the rapid loss of trout, salmon, pike, and bass. Spraying crops for insect pests had inadvertently killed off many predaceous mites, resulting in outbreaks of herbivorous mites that obviously suffered less. Forest spraying has similarly 'released' populations of scale insects after heavy damage to their wasp enemies.

In addition, SCEP points out that

such fat-soluble pesticides as DDT are concentrated as they pass from one feeding level to the next. In the course of digestion a predator retains rather than eliminates the DDT content of its prey. The more it eats, the more DDT it accumulates. The process results in especially high concentrations of toxins in predaceous terrestrial vertebrates.

Predators also suffer from the destruction of their food supply. Severe damage to the lower levels in the food chain usually leads to the extinction of the predator before that of the species on which it preys.

There is yet another way in which we are reducing complexity. Populations at any given moment will be made up of individuals of every possible age group. We tend to replace such balanced populations with plantations of trees and other crops which are all of the same age and are particularly vulnerable to diseases affecting them at particular stages in their life cycle. This principle must apply equally well to intensive stock-rearing units, and especially factory farms. Once more the result is to reduce stability.

Technological devices must also reduce complexity. They constitute external controls exerted by precarious human manipulation. They invariably replace natural controls of a far more complex nature. Thus, to replace the natural controls which ensure the stability of an insect population by a single chemical pesticide involves a drastic reduction in complexity. The same must be true when we replace the natural mechanisms ensuring soil fertility with nitrogen, phosphorus and potassium, which are the main ingredients of artificial fertilizers.

In fact, most human activities are reducing the stability of the ecosphere, which is simply another way of saying that they are determining its systematic degradation.

For several thousand million years, the ecosphere has been developing into an extremely complex organization of different forms of life in close interaction with each other. In doing this it has been counteracting the basic tendency of all systems towards randomness or entropy. The elaborate mechanisms that have enabled the ecosphere to develop in this manner have been disrupted by man's activities. In his gross presumption, he has sought to replace them with devices causing dereliction and confusion, which, rather than seek to satisfy the countless competing requirements of the ecosphere, have been geared to the satisfaction of petty, short-term anthropocentric ends. As a result, the organizational process has been reversed; waste, or random parts, are accumulating faster than organization is building up. Rather than counteract the inexorable trend towards entropy, industrial man's activities are accelerating it.

If these activities continue to increase exponentially at 6·5 per cent per annum, or double every 13½ years, it cannot take many decades before our planet becomes incapable of supporting complex forms of life.

POLLUTION

Studies of the effects of pollutants on ecosystems have often yielded contradictory results. Rather than attempt to weigh these up, we have chosen to summarize some of the findings of what is almost certainly the most authoritative study, that undertaken in 1969 by an impressive group of scientists from many different disciplines under the auspices of MIT and referred to as the Study of Critical Environmental Problems or SCEP. This study is to be used as background material for the UN Conference on the Human Environment 1972.

SCEP accentuates the necessity for adopting a holistic approach.

The significant aspect of human action is man's total impact on ecological systems, not the particular contributions that arise from specific pollutants. Interaction among pollutants is more often present than absent. Furthermore, the total effect of a large number of minor pollutants may be as great as that of one major pollutant. Thus, the total pollution burden may be impossible to estimate except by direct observation of its overall effect on ecosystems.

The scale of human activity can be estimated by comparing specific man-induced processes with the natural rates of geological and ecological processes. It can be shown that in at least twelve cases man-induced rates are as large or larger than the natural rates (see Table 1).

It is pointed out that with a five per cent natural growth increment in the mining industries, this will apply to many more materials.

... these comparisons show that at least some of our actions are large enough to alter the distribution of materials in the biosphere. Whether these changes are problems depends upon the toxicity of the material, its distribution in space and time, and its persistence in ecological terms.

Most of the disruptive processes already described are well advanced, however, and as they occur slowly the most visible effect is a gradual deterioration of ecosystems, 'characterized by instability and species loss'.

Many lakes and urban centres have severely deteriorated ecosystems. Less severe deteriorations occur more commonly, often as temporary afflictions in ecosystems that otherwise manage to survive intact. This general problem is labelled 'attrition' because it lacks discrete steps of change. Stability is lost more and more frequently, noxious organisms become more common,

Table 1   Man-induced rates of mobilization of materials which exceed geological rates as estimated in annual river discharge to the oceans.

(Thousands of metric tons per year).

Element	Geological Rates* (in rivers)	Man-Induced Rates† (mining)
Iron	25,000	319,000
Nitrogen	8,500	9,800 (consumption)
Manganese	440	1,600
Copper	375	4,460
Zinc	370	3,930
Nickel	300	358
Lead	180	2,330
Phosphorus	180	6,500 (consumption)
Molybdenum	13	57
Silver	5	7
Mercury	3	7
Tin	1·5	166
Antimony	1·3	40

Reproduced from SCEP.

  * Source: Bowen, 1966.

  † Source: United Nations, *Statistical Yearbook*, 1967. Data for mining except where noted.

and the aesthetic aspects of waters and countryside become less pleasing. This process has already occurred many times in local areas. If it were to happen gradually on a global scale, it might be much less noticeable, since there would be no surrounding ecosystems against which to measure such slow changes. Each succeeding generation would accept the *status quo* as 'natural'.

ENERGY PRODUCTS

Present and future levels of energy consumption are particularly relevant to estimating our capacity to disrupt ecosystems. The best available calculation appears to be that made by the Battelle Memorial Institute in 1969. In 1968 energy consumption in the US was slightly over 60,000 trillion BTU. It appears to be rising at 3·2 per cent per annum and is expected to be 170,000 trillion BTUs by the year 2000.

Over the last 50 years there has been a decreasing amount of energy used for each unit of GNP. The increased technical efficiency of energy used has tended to more than offset the more intense use of energy. The trend, however, appears to be changing. The present policy is to encourage energy use while the technical efficiency of new electric-power plants and other energy-conversion devices is no longer increasing and may even decrease over the next decades. If this is so, then it is possible that this and other projections have underrated future energy requirements. On the other hand conservation pressures might lead to a reduced usage and this has not been taken into account.

World-wide energy consumption projection made by Joel Darmstadter of Resources for the Future has appeared in a work *Energy and the World Economy* (see Table 2).

What are likely to be the emissions from power production and other forms of energy production?

It is estimated that in 1967 some 13·4 billion metric tons of CO_2 were released from fossil fuel combustion and that emissions in 1980 (using Darmstadter's projection) would be 26 billion metric tons for the world as a whole.

SCEP points out that the trend towards depleting the remaining stands of original forests, such as those in tropical Brazil, Indonesia and the Congo, will further reduce the capacity of the ecosphere to absorb CO_2 and may release even more CO_2 to the atmosphere. The CO_2 content of the atmosphere is increasing at a rate of 0·2 per cent per year since 1958. One can project, on the basis of these trends, an 18 per cent increase by the year 2000, i.e. from 320 ppmm to 379 ppmm. SCEP considers that this might increase temperature of the earth by 0·5°C. A doubling of CO_2 might increase mean annual surface temperatures by 2°C. (see Table 3).

HEAT

Thermal waste energy is increasing at a rate of 5·7 per cent per annum, which means that it is likely to increase by a factor of 6

Table 2   Darmstadter's projection of world energy consumption

	Solid			Liquid	
	A*	10^{12} kWh(t)‡	Percentage of World Consumption	10^{12} kWh(t)†	Percentage of World Consumption
Developed Countries					
United States	3·5	5·0	17·3	9·4	25·3
Canada	5·5	0·3	0·9	1·2	3·2
Western Europe	4·0	2·7	9·4	9·2	24·9
Communist Eastern Europe	4·6	3·6	12·5	1·5	3·9
USSR	6·5	5·7	19·7	5·2	14·0
Japan	7·9	0·5	1·9	3·5	9·5
Oceania	4·8	0·4	1·3	0·4	1·2
Total	4·7	18·2	63·0	30·4	82·0
Developing Countries					
Communist Asia	7·6	7·3	25·4	0·7	2·0
Other Asia (exc. Japan)	8·5	2·3	3·1	2·4	6·5
Africa	6·5	0·9	0·6	0·7	1·9
Other America	7·4	0·2	7·9	2·8	7·6
Total	7·7	10·7	37·0	6·6	18·0
World Total	5·2	28·9	100·0	37·0	100·0

Source: Estimated by Joel Darmstadter in *Energy and the World Economy* (to be published
  * Column A contains the projected average annual percentage of growth in energy
  † Darmstadter follows the UN system of evaluating hydro and nuclear electricity. This
for kWh per 0·125 m.t.c.e. (for the factor see UN *World Energy Supplies* or the Appendix
  ‡ Converted from metric tons coal equivalent by using 27·3 × 10⁶ Btu/m.t.c.e. and
  § Unknown, but believed to be small.

before the end of the century. The total for 1970 was $5.5 \ 10_9$ MW, which is likely to increase to 9·6 by 1980 and $31.8 \ 10^6$ MW by 2000. The effects on global climate are not known.

Emissions of pollutants such as sulphur oxides, nitrogen oxides, hydrocarbons, carbon monoxide and particulate matter cannot be predicted with any assurance. The theoretical knowledge necessary to make these predictions does not yet exist nor are the relevant facts available.

As far as emissions of radionuclides are concerned, the major source will be at the site of fuel-reprocessing plants. One estimate is that 99·9 per cent of all such emissions entering the environment are from such sources. Concern is expressed for emissions of 'potentially hazardous' radionuclides such as iodine-131, xenon-153, strontium-90, and caesium-137. Possible releases of tritium (hydrogen-3) and krypton-85 are also of concern.

in 1980.

Gas		Hydro†		Nuclear†		Overall	
10^{12} kWh(t)‡	Percentage of World Consumption	10^{12} kWh(t)‡	Percentage of World Consumption	10^{12} kWh(t)‡	Percentage of World Consumption	10^{12} kWh(t)‡	Percentage of World Consumption
8·3	41·9	0·34	18·1	0·98	52·0	24·0	26·8
1·0	4·8	0·22	11·6	0·05	2·8	2·8	3·0
2·1	10·3	0·46	24·1	0·63	33·5	15·1	16·8
0·7	3·5	0·02	1·2	0·04	2·1	5·9	6·6
5·9	29·8	0·29	15·3	0·04	2·1	17·1	19·1
0·1	0·4	0·11	5·7	0·11	5·6	4·3	4·9
0·1	0·7	0·04	2·1	0·01	0·3	1·0	1·1
18·2	91·4	1·48	78·1	1·86	98·4	70·2	78·3
—	—§	0·04	2·2	0·01	0·4	8·1	9·1
0·4	2·2	0·14	7·3	0·02	1·0	5·2	5·8
0·2	0·9	0·06	3·0	—	—	1·8	2·0
1·1	5·5	0·18	9·4	0·00	0·2	4·3	4·8
1·7	8·6	0·42	21·9	0·03	1·6	19·4	21·7
19·9	100·0	1·90	100·0	1·89	100·0	89·6	100·0

by The Johns Hopkins Press for Resources for the Future, Inc).
consumption for 1965–80.
means that he used for *both* nuclear and hydro-power the system used by the Group *only* of any recent U N *Statistical Yearbook*).
0.293×10^{-3} kWh(t)/Btu.

Total emissions would not lead to anything like maximum permissible concentrations (MPC) if dispersal was assured. However, one must take into account the tendency of radionuclides to concentrate in certain organisms and to get into food chains. Concentration factors of 1,000 for caesium in the flesh of bass have been found, of 8,700 in the bones of the blue gills, of 350,000 for radioactivity content in caddis-fly larvae, 40,000 for duck egg yolks and 75,000 for adult swallows. Table 4 shows estimated concentration factors for some radionuclides in aquatic organisms.

Table 3 CO_2 produced by fossil fuel combustion, 1950–67.
(Billions of metric tons) Reproduced from SCEP.

Year	Coal	Lignite	Refined Oil Fuels	Natural Gas	Total
1950	3·7	0·9	1·4	0·4	6·4
1951	3·8	0·9	1·7	0·5	6·9
1952	3·8	0·9	1·8	0·5	7·0
1953	3·8	0·9	1·9	0·5	7·1
1954	3·8	0·9	2·0	0·6	7·3
1955	4·1	1·0	2·2	0·6	7·9
1956	4·4	1·1	2·4	0·7	8·6
1957	4·5	1·3	2·5	0·7	9·0
1958	4·6	1·4	2·6	0·8	9·4
1959	4·8	1·4	2·8	0·9	9·9
1960	5·0	1·4	3·1	1·0	10·5
1961	4·5	1·5	3·3	1·0	10·3
1962	4·6	1·5	3·5	1·1	10·7
1963	4·8	1·6	3·8	1·2	11·4
1964	5·0	1·7	4·2	1·3	12·2
1965	5·0	1·7	4·5	1·5	12·7
1966	5·1	1·7	4·8	1·6	13·2
1967	4·8	1·7	5·2	1·7	13·4
1980 (est.)	11·1		10·8	4·0	26·0

Phytoplankton also tend to concentrate activation products such as zinc-65, cobalt-67, iron-55 and manganese-54 to an even greater extent than fission products.

Table 4 Estimated concentration factors in aquatic organisms.

Radionuclide	Site	Phytoplankton	Filamentous Algae	Insect Larvae	Fish
Na^{24}	Columbia River	500	500	100	100
Cu^{64}	Columbia River	2,000	500	500	50
Rare earths	Columbia River	1,000	500	200	100
Fe^{59}	Columbia River	200,000	100,000	100,000	10,000
P^{32}	Columbia River	200,000	100,000	100,000	100,000
P^{32}	White Oak Lake	150,000	850,000	100,000	30–70,000
Sr^{90}–Y^{90}	White Oak Lake	75,000	500,000	100,000	20–30,000

Reproduced from SCEP.
Source: Eisenbud, 1963.

When breeder reactors are introduced plutonium emissions will also become a concern.

The management of concentrated and highly radioactive wastes is a serious problem deserving far more study. Table 5 provides an estimate of accumulated wastes for 1970, 1980 and 2000.

DOMESTIC AND AGRICULTURAL WASTES

Dredged wastes from urban areas contain sediment, sewage solids, agricultural and industrial wastes. These also tend to be deposited in rivers or coastal waters. The total amount deposited in this way is estimated at between 150 and 220 million metric tons per years, and appears to be increasing at 4 per cent per annum.

Table 5 Radioactive wastes as a function of expanding US nuclear power.

	1970	1980	2000
Installed nuclear capacity, MW(e)	11,000	95,000	734,000
Volume high-level liquid waste*†			
Annual production, gal/yr	23,000	510,000	3,400,000
Accumulated volume, gal‡	45,000	2,400,000	39,000,000
Accumulated fission products, megacuries†			
Sr^{90}	15	750	10,800
Kr^{85}	1·2	90	1,160
H^3	0·04	3	36
Total for all fission products	1,200	44,000	860,000
Accumulated fission products, tons	16	388	5,350

Source: Snow, 1967 (reproduced from SCEP).

*Based on 100 gallons of high-level acid waste per 10,000 thermal megawatt days (MWd) irradiation.

†Assumes 3-year lag between dates of power generation and waste production.

‡Assumes wastes all accumulated as liquids.

World production and consumption of chemical fertilizers (except during periods 1914–18 and 1940–5) have doubled or tripled in each decade. Total world use in 1963–4 exceeded 33 million metric tons, only 10 per cent of which were used in developing countries. Their share, however, is increasing rapidly.

Present annual world production of pesticides is probably about 1 million metric tons. It is likely to go on increasing in view of the increasing world food shortage and because of diminishing returns on their use. Thus to double world food production, which as we have seen is likely to be necessary, it will be necessary to increase consumption by no less than six times (see Table 6).

In the industrialized countries there is likely to be a move away from DDT to less persistent but more toxic pesticides such as phorate, dimiton, parathion, etc. These require more frequent sprayings to make up for their reduced persistence. It is unlikely that the developing countries will be able to afford

Table 6 Pesticides needed to increase food production on acreage now under cultivation in Asia (except mainland China and Japan), Africa, and Latin America by the percentages indicated.

Percentage of Increase in Agricultural Production	Tonnage Needed (metric tons)
—	120,000
10	150,000
20	195,000
30	240,000
40	285,000
50	342,000
60	402,000
70	475,000
80	558,000
90	640,000
100	720,000

Reproduced from SCEP.
Source: President's Science Advisory Committee (PSAC), 1967.

them, so consumption of DDT is likely to continue growing.

SCEP points out the way in which agriculture becomes increasingly dependent on the use of these poisons:

Realisation that the use of pesticides increases the need to continue their use is not new, nor is the awareness that the constant use of pesticides creates new pests. For many of our crops on which pesticide use is heavy, the number of pests requiring control increases through time. In a very real sense, new herbivorous insects find shelter among our crops where their predator enemies cannot survive. Fifty years ago most insect pests were exotic species, accidently imported to a country lacking their natural enemies. More recently many of the pests, including especially the mites, leaf-rolling insects, and a variety of aphids and scale insects, have been indigenous. Thus pesticides not only create the demand for future use (addiction), they also create the demand to use more pesticide more often (habituation). Our agricultural system is already heavily locked into this process, and it is now spreading to the developing countries. It is also spreading into forest management. Pesticides are becoming increasingly 'necessary' in more and more places. Before the entire biosphere is 'hooked' on pesticides, an alternative means of coping with pests should be developed.

Of all pesticides, DDT is the most commonly used, and is now present in the fatty tissue of animals in every part of the world. Its effects are well documented. SCEP summarizes some of the implications:

The oceans are an ultimate accumulation site of DDT and its residues. As much as 25 per cent of the DDT compounds produced to date may have been transferred to the sea. The amount in the marine biota is estimated to be in the order of less than 0·1 per cent of total production and has already produced a demonstrable impact upon the marine environment.

Population of fish-eating birds have experienced reproductive failures and population declines, and with continued accumulation of DDT and its residues in the marine ecosystem additional species will be threatened. The decline in productivity of marine food fish and the accumulation of levels of DDT in their tissues can only be accelerated by DDT's continued release to the environment.

Certain risks in the utilisation of DDT are especially difficult to quantify, but they require most serious consideration. The rate at which it degrades to harmless products in the marine system is unknown. For some of its degradation products, half-lives are certainly of the order of years, perhaps even of decades. If most of the remaining DDT residues are presently in reservoirs which will in time transfer their contents to the sea, we may expect, quite independent of future manufacturing practices, an increased level of these substances in marine organisms. And if, in fact, these compounds degrade with half-lives of decades, there may be no opportunity to redress the consequences. The more the problems are studied, the more unexpected effects are identified. In view of the findings of the past decade, our prediction of the hazards may be vastly underestimated.

HEAVY METALS

Pollution by heavy metals also gives cause for concern.

Some heavy metals are highly toxic to plants and animals including man. They are highly persistent and retain their toxicity for very long periods of time. Some have been used extensively as pesticides and have been dispersed into the environment as pesticides, as uncontrolled industrial wastes and emissions and other means.

Much enters natural water systems through sewage discharges and only a portion is removed by normal sewage treatment.

Those heavy metals that are most toxic, persistent and abundant in the environment have been selected by SCEP for special review. These include mercury (Hg), lead (Pb), arsenic (As), cadmium (Cd), chromium (Cr), and nickel (Ni). Most heavy metals are biologically accumulated in the bodies of organisms, remain for long periods of time, and function as cumulative poisons. Table 7 indicates world production of these metals between the years 1963 and 1968 and illustrates the rate at which it is increasing.

It may be worth looking more closely at the problem of mercury pollution, which is particularly topical. SCEP quotes Stockinger:

Elemental mercury and most compounds of mercury are protoplasmic poisons and therefore may be lethal to all forms of living matter. In general, the organic mercury compounds are more toxic than mercury vapour or the inorganic compounds. Even small amounts of mercury vapour or many mercury compounds can produce mercury intoxication when inhaled by man. Acute mercury poisoning, which can be fatal or cause permanent damage to the nervous system, has resulted from inhalation of 1,200 to 8,500 micrograms per cubic meter of mercury. The more common chronic poisoning (mercurialism) which also affects the nervous system is an insidious form in which the patient may exhibit no well-defined symptoms for months or sometimes years after exposure.

Mercury is also dangerous when ingested in food. In Japan 111 cases of mercury poisoning occurred (with 44 deaths) as a result of eating fish taken from Minamata Bay. Another outbreak occurred at Big Niigata City with 26 cases (and five deaths).

Mercury's toxicity is permanent. In addition, when fish, shellfish, birds or mammals containing mercury are eaten by other animals the mercury may be absorbed and accumulated.

Industrial wastes and agricultural pesticides have caused severe mercury contamination in waters in Japan, Sweden and the US. Its use is increasing throughout the world and it 'threatens to become critical in the world environment'. Moreover, as SCEP points out, mercury is but one of approximately two dozen metals that are highly toxic to plants and animals.

OIL POLLUTION

We tend to regard oil pollution of the seas as caused principally by accidental spills like that of the Torrey Canyon. Such accidents cause the most evident damage, but they make up less than 10 per cent of the estimated 2·1 million metric tons of oil that man introduces directly into the world's waters. At least 90 per cent originates in the normal operations of tankers, other ships, refineries, petro-chemical plants, and submarine oil-wells; from disposal of spent lubricants and other industrial and automotive oils; and by fall-out of airborne hydrocarbons emitted by vehicles and industry (see Table 8).

The actual amount that goes directly into the seas must be taken as proportionate to production. It is normally estimated at 0·1 per cent of production but if possible fall-out of airborne hydrocarbons on the sea surface is added it may be as much as 0·5 per cent. This is because estimated emissions of hydro-

Table 7 World production * and US consumption† of toxic heavy metals. *(Thousands of metric tons).*

Year	Hg World	Hg US	Cd World	Cd US	Pb World	Pb US	Cr₂O₃ World	Cr₂O₃ US	Ni World	Ni US
1960	—	1·77	—	4·53	—	930	—	1,110	—	98·2
1961	—	1·92	—	4·65	—	932	—	1,090	—	108
1962	—	2·26	—	5·56	—	1,010	—	1,030	—	108
1963	8·28	2·70	11·8	5·19	2,520	1,060	3,920	1,080	340	114
1964	8·81	2·81	12·7	4·31	2,520	1,090	4,150	1,320	372	134
1965	9·24	2·54	11·9	4·75	2,700	1,130	4,810	1,440	425	156
1966	9·51	2·46	13·0	6·60	2,860	1,200	4,390	1,330	414	171
1967	8·36	2·40	12·9	5·28	2,880	1,150	4,300	1,230	441	158
1968	8·81	2·60	14·1	6·05	3,000	1,200	4,730	1,200	480	144‡

Reproduced from SCEP.

* Sources: 1963 data are from the *Minerals Yearbook*, 1967; 1964–8 data are from the *Minerals Yearbook*, 1968.

† Source: *Chemical Economics Handbook*, 1969.

‡ Source: *Minerals Yearbook*, 1968.

carbons of petroleum origin to the air is 90 million tons, 40 times that emitted to the seas. Nobody knows how much may finally settle in the seas. SCEP points out that, if 'io per cent does, then the total hydrocarbon contamination of the oceans could be almost five times the direct influx from ships and land sources.'

The increase in the size of tankers must make things worse. The danger of large-scale accidents will increase with the scale of the tankers. 800,000-ton tankers are projected. 'A single spill from one of these would add 20 per cent to the amount of oil entering oceans in a single year' (SCEP). Cleaning up oil spills does more harm than good 'even with a non-toxic dispersant, the dispersed oil is much more toxic to marine life than is an oil slick on the surface' (SCEP).

The effect of spills in shallow water is particularly damaging. Thus 'an accidental release of 240 to 280 tons of No. 2 fuel oil from a wrecked barge off West Falmouth, Massachusetts, in 1969 caused an immediate massive kill of organisms of all kinds – lobsters, fish, marine worms and molluscs.'

Table 8   Estimates of direct losses into the world's waters, 1969.

(*Metric tons per year*).

	Loss	Percentage of Total Loss
Tankers (normal operations)		
Controlled	30,000	1·4
Uncontrolled	500,000	24·0
Other ships (bilges, etc.)	500,000	24·0
Offshore production (normal operations)	100,000	4·8
Accidental spills		
Ships	100,000	4·8
Non-ships	100,000	4·8
Refineries	300,000	14·4
In rivers carrying industrial automobile		
wastes	450,000	21·6
Total	2,080,000	100·0

Reproduced from SCEP.

The difficulty of estimating biological effects in coastal waters is that 'many other pollutants are also present in this zone and it is hard to separate their different effects. Indeed, the effects may not be separable, but instead additive or mutually reinforcing'.

One possible effect of oil dispersed over wide ocean areas could arise from the fact that

chlorinated hydrocarbons such as DDT and Dieldrin are highly soluble in oil film. Measurements ... in Biscayne Bay, Florida, showed that the concentration of a single chlorinated hydrocarbon (dieldrin) in the top 1 millimetre of water containing the slick was more than 10,000 times higher than in the underlying water ... We know that the small larval states of fishes and both the plant and animal plankton in the food chain tend to spend part of the night hours quite near the surface, and it is highly probable that they will extract, and concentrate still further, the chlorinated hydrocarbons present in the surface layer. This could have seriously detrimental effects on these organisms and their predators.

Implicit throughout this study is the knowledge that these ecologically disruptive trends cannot be allowed to persist indefinitely. SCEP concludes:

In general, the expected losses from present impacts do not exceed our capacity to carry the burden; this leads us to the conclusion that an intractable crisis does not now seem to exist. Our growth rate, however, is frightening. The impact of two, four, or eight times the present ecological demand will certainly incur greater losses in the environment. If the process of change were gradual, the present ecological advantage that is reflected in our 5 to 6 per cent annual growth would taper off in the face of decreased environmental services, and growth would be correspondingly slowed. Instead, the risk is very great that we shall overshoot in our environmental demands (as some ecologists claim we have already done), leading to cumulative collapse of our civilisation. It seems obvious that before the end of the century we must accomplish basic changes in our relations with ourselves and with nature. If this is to be done, we must begin now. A change system with a time lag of ten years can be disastrously ineffectual in a growth system that doubles in less than fifteen years.

Appendix B:
Social systems and their disruption

The activities of industrial man are having a very serious effect on society. They can be shown to be leading to its disintegration, and it can also be shown that such pathological manifestations as crime, delinquency, drug addiction, alcoholism, mental diseases, suicide, all of which are increasing exponentially in our major cities, are the symptoms of this disintegration.

Unfortunately, before we can understand why and how this is happening, we must know a little more about human society. Sociology, which should provide us with this information, is failing to do so, mainly because it is studying human society *in vacuo*, i.e. without reference to behaviour at other levels of organization. This is the result of regarding man and the societies he develops as unique, and in some way exempt from the laws governing all the other parts of the ecosphere. If we establish this false dichotomy between man and other animals it is partly because we fail to understand the nature of the evolutionary process. Thus, owing to our tendency towards subjective classification, we recognize that certain events among which a connection can be made within our immediate experience can be regarded as constituting one process, while, on the other hand, we refuse to admit that this can be the case with events whose connecting bond lies outside our experience. Thus we are willing to admit that the development of a foetus into an adult is a single process, and that it is difficult to examine, separately and in isolation, any of its particular stages apart from the process as a whole. On the other hand, we are less ready to regard evolution in this way.

We still imply that radical frontiers exist between life at different levels of complexity, in spite of the fact that they are part of the same evolutionary process. Yet, it can be demonstrated that no such frontiers obtain. When Kohler synthesized urea, the barrier between the 'organic' and the 'inorganic' was suddenly shattered, as was that between the 'animate' and 'inanimate' when the virus was found to manifest certain features associated with life on being confronted with a source of protein, and at other periods to display the normal behaviour pattern of a crystal. Again, it has been demonstrated repeatedly that no barrier exists separating man from other animals. He is more 'intelligent', and that is about all that can be said.

If human societies are not unique, their functions cannot be understood apart from that of other natural systems, such as ecosystems and biological organisms, i.e. in the light of a general theory of behaviour.

To understand this, one must first realize that the vast and chaotic human societies in which we are living are by no means normal. If man has been on this planet for a million and a half years, which is possible, it is only in the last 150 years that he has become an industrialist, and that industry has permitted the development of such societies. This represents no more than two days in the life of a man of 50.

For at least a million years and probably more he earned his living as a hunter-gatherer. During all this time, there is no reason to suppose that the societies he developed were in any way less adapted to their respective environments than are those of non-human animals.

From our knowledge of surviving hunter-gatherer societies, such as the Bushmen of the Kalahari, one can presume that they probably consumed less than a third of the available food resources. They did not clear forests for agricultural land, nor did they hack down trees for building houses, nor were they so short-sighted as to exterminate the wild animals on which they depended for their livelihood.

At the same time they avoided increasing their population over and above that which might lead them to have to alter their life-style in any way. Even if one considers an area overpopulated, as does Professor Ehrlich (1), 'when human numbers are pressing against human values', and not just when they actually starve, then such societies were never overpopulated.

What is more, the survival of such societies was compatible with that of climax ecosystems, to which they contributed by fulfilling within them their various ecological functions. Take the case of the Plains Indians of North America, who lived off the vast herds of bison. They did not, on the whole, attack the main herd, which would have been a dangerous undertaking, but rather killed off the stragglers, the old and the weak, thereby exerting quantitative as well as qualitative controls on these animals. It is significant that exactly the same is true of the lions living off the buffalo herds of East Africa.

If human societies for 99·75 per cent of their tenancy of this planet behaved as an integral part of our ecosphere (before the invention of agriculture 10,000 years ago and industry 150 years ago) it is unreasonable to suppose that such behaviour is not subject to its laws.

Nor is there any reason why sociology should be anything but a branch of the natural sciences, that which deals with a particular type of natural system: the human society. Let us briefly look at human society in this light.

First of all, like all other natural systems, a human society displays organization. This is probably its most important feature. If one gathers on an island a random collection of people from different societies speaking different languages it would be naïve to suggest that these constituted a society. Nevertheless there would be a tendency for organization, or negative entropy, to build up (or entropy and randomness to be reduced). First of all men would pair off with women and have children. Families would be formed and groups of these families would tend to be associated and grow into small communities. As this occurred so their members would develop more and more things in common. They would learn to speak the same language and dress, eat and build their houses in a similar way. Slowly a common set of values and aspirations would emerge, and these would bind them together in a common purpose and transform them into a true society.

This organizational process is not a linear one. Thus, in its development from the simple to the complex, matter passes through certain critical stages, where the possibilities of a particular type or organization are exhausted and further advance can only be achieved by the development of a new type. Thus, an atom can be developed only up to a certain point. This point will vary with different types of atoms, some of which, such as the tungsten atom, are relatively large. Beyond this critical point, however, development can occur only by the association of several atoms together to form a molecule. As soon as the latter stage is reached, the constituent atoms undergo a considerable change, in that a radical division of labour occurs, in accordance with the law of economy. To explain their behaviour now requires the introduction of new principles.

There is no reason to suppose that this notion of levels of organization does not apply equally well to human social systems. Thus the family, which clearly represents the first level of human organization, is a universal feature of all human societies, and there is no example of its suppression without the most serious social consequences. The family is held together by bonds which are extendable in the sense that the stimuli required for triggering off the corresponding behavioural responses are not specific as in the case of simpler forms of life.

For example, not only a mother but a mother-like figure can trigger off filial responses, or vice versa (2). It is this feature of the family bonds which permits the development of larger social units. The latter can, of course, be of many kinds. They can be bilateral extended families, or unilateral, or the members of the different families constituting these units need not be related at all, as mere contiguity is sufficient to allow the development of such bonds (3).

Another essential characteristic of the family bonds is that they cannot be extended indefinitely. This is a feature of all bonds, whether they be holding together the nucleus of an atom or the solar system. A point must, therefore, be reached where the bonds cannot be extended any further, and development becomes possible only by the association of a number of such units. At this point it can be said to have reached a new level of organization.

Once we pass the level of the village, clan or lineage, we reach a level of social organization that has not often been achieved by the human species. To harness the family bonds in such a way as to build up a larger unit requires the development of very elaborate forms of organization. This involves 'criss-cross' bonds that permit the establishment of a veritable cobweb of associations of one sort or another, all of which transcend each other in such a way that each individual is linked to each other member of the society in at least one, and preferably more ways.

Thus a tribesman is at once a member of a family and of a maternal and of a paternal kinship group. As none of these may coincide with the social unit that is the village in which he lives, he is a member of yet another group: the village. He is also likely to be a member of an age grade, of a secret society of some sort, possibly also of a military club, and of some other group with a common economic activity. Such a man has a very definite status, which Linton (4) defines as 'the sum total of all the statuses which he occupies and hence his position with relation to the total society'.

The same principle is apparent in the more stable segment of our modern societies. As Linton writes: 'The status of Mr Jones as a member of his community derives from a combination of all the statuses which he holds as a citizen, as an attorney, as a Mason, as a Methodist, as Mrs Jones's husband, and so on.' As a result of such criss-cross associations, a man is in contact with a very large number of cross-sections of the society. There is what Ortega y Gasset (5) calls 'social elasticity'.

All the parts of the society are in contact with each other. Any change in the society will, therefore, effect each individual, and the actions of each individual must effect the society as a whole through the agency of all the associations of which he is a member.

Without social elasticity there would be no bonds, no organization: in fact no real society. Yet social elasticity can be maintained only in special conditions. Thus it is likely that if the society grows too big the bonds holding it together become of an ever more precarious nature and eventually incapable of holding it together.

The social system is, in fact, overloaded, with more people

than it is capable of organizing into a society. Its essential structure breaks down, and it ceases to be capable of self-regulation.

As already mentioned, it is a basic feature of all bonds that there is a limit to their extendability. Those holding together a community which are already extensions of the family ones cannot be extended to hold together more than a certain number of people. Aristotle considered that a city could be made up of no more citizens than could know each other by sight. The Greek city-states, which displayed some of the features of self-regulating units, were, in fact, very small. Only three had more than 20,000 citizens (Athens, Corinth and Syracuse). It is significant that a recent study in America has revealed that the crime rate appears to be proportionate to the size of the city. Violent crime appears to be four times greater *per capita* in cities of 250,000 people than in cities of 10,000 (see Table 8). Unfortunately it is one of the essential features of industrial society that it gives rise to urbanization (see Figure X, World Urban Population, p. 26 in *The Limits of Growth*).

'Social elasticity' is also seriously affected by mobility. It is impossible to create sound societies when people are being constantly moved from place to place. In such conditions, the towns are not made up of people who have grown up together and among whom bonds have had time to develop, but simply of people who have been thrown together for various random reasons. Bonds cannot be manufactured at will. Nor can that socialization process that will enable people to fulfil their specific functions within their social system be compressed into a few years of adult life. It is a slow, educative process, the most important part of which must occur in the early years of life – when the generalities of a cultural pattern, i.e. its basic goals and values, are inculcated via the family and the small community.

To understand this principle, it is necessary to see how cultural information is used to determine the adaptive and self-regulatory behaviour of a social system, in fact, how the basic cybernetic model applies to a society.

If a society is capable of self-regulating behaviour, it is that its responses are based on a model of its relationship with its environment, in the light of which they are being continually monitored. Such a model is a society's world-view, or *Weltan-schauung*, which is compounded of its religion, mythology, traditional law, and so on.

As soon as one understands a society's culture, one understands the reason for its behaviour, and all its actions that previously appeared random or irrational now appear quite logical. The following example illustrates this point.

It is well known that some Australian aborigines failed to establish a cause and effect relationship between copulation and conception. Instead, they generally believed that the spirits of chilren yet unborn, which were apparently referred to as 'ngargugalla', inhabited some strange world from which they emerged only when dreamt of by their mothers. Daisy Bates (6) tells us that among the Koolarrabulloo it was the father who had to have such a dream:

> They believed that below the surface of the ground and at the bottom of the sea was a country called Jimbin, home of the spirit babies of the unborn, and the young of all the totems. In Jimbin there was never a shadow of trouble or strife or toil or death: only the happy laughter of the little people at play. Sometimes these spirit babies were to be seen by the jalngangooroo, the witch-doctors, in the dancing spray and sunlight of the beaches, under the guardianship of old Koolibal, the mother-turtle, or tumbling and somersaulting in the blue waters with Pajjalburra, the porpoise . . .
>
> So firm was the belief in the 'ngargalulla' that no man who had not seen it in his sleeping hours would claim the paternity of a child born to him. In one case, that came under my observation, a man who had been absent for nearly five years in Perth proudly acknowledged a child born in his absence, because he had seen the 'ngargalulla', and, in another, though husband and wife had been separated not a day, the man refused absolutely to admit paternity. He had not dreamed the 'ngargalulla'. Should a boy arrive when a girl came in the dream, or should the ngargalulla not have appeared to its rightful father, the mother must find the man who has dreamed it correctly, and he is ever after deemed to be the father of that child.

It is evident that, if we were not aware of this aspect of the world-view or model of the Koolarrabulloo, we would find their attitude towards the acceptance of paternity totally illogical. However, once we were acquainted with their model, their attitude would appear quite reasonable, and could even be predicted with a fair measure of probability. There is no reason why all seemingly irrational behaviour cannot be explained on the basis of a cultural model of this sort.

We regard as 'rational', behaviour which is based on *our* cultural model of the world, and which somewhat presumptuously we consider to be the only valid one. Indeed, 'rational behaviour', if it is to have any meaning, must be taken as based on a model, as opposed to 'irrational' or random behaviour, which is not. In this case the behaviour of even the most primitive societies must be regarded as rational. However, if we look at things holistically and functionally it becomes apparent that this is not so. Some may object that the model in question must be a 'valid' one, by which it is presumably meant that it must be scientific rather than derived from cultural tradition.

Society is limited in its capacity to adapt to environmental change, not only by its inability to comprehend it, but also, among other things, by the social disruption that must be caused by attempting to adapt too rapidly to such change. Science may facilitate comprehension, and may provide the technology for ensuring partial short-term adaptation to change, but, by failing to provide the means for ensuring the corresponding adaptation of social structures, it must thereby determine their disintegration.

There is every reason to suppose that a traditional culture permits that interpretation of the society's relationship with its environment that is consistent, among other things, with the maintenance of social structures. Indeed, this appears in fact to be the principal, culturally induced, preoccupation of a traditional society, to which all others are subordinated. This, by necessity, must mean opposing change to which social adaptation is not possible, rather than attempting to adapt to it on insufficient information, as is done in our science-based society. If, as we must suppose, the role of a cultural behaviour, like that of any other type of behaviour, is to ensure the stable relationship of a system, in this case a social one, with its environment, then we must accept that the cultural behaviour of the traditional societies that we derogatorily describe as primitive is more 'valid' than our own.

THE GOAL

Equally important is the fact that a culture also provides a society with a goal-structure and a means of achieving it. The goal of all self-regulating societies appears to be the acquisition of social prestige. It is important to realize that this goal is possible only in a closely knit society in which there is fundamental agreement as to what are the determinants of prestige. These will vary in each society. In general one can say that these will coincide with the qualities that must be cultivated if the society is to survive. Thus in a society of hunter-gatherers, success in a hunt is likely to be a determinant of prestige; among societies involved in war-like pursuits courage is likely to be particularly prestigious. The prestige achieved will determine one's position in the social hierarchy. This hierarchy is of immense importance in avoiding strife and in ensuring a socially acceptable division of labour among the members of the society. If there is no hierarchy there will be constant bickering and fighting. There will also be no mechanism for ensuring the perpetuation of those qualities required if the society is to survive.

Hierarchy is another word for organization. There are only two ways of dispensing with it: one is to accept chaos and with it asystemic controls such as dictators; the other is to reduce the size of the society. In an extremely small social grouping such as the Kalahari Bushmen and the Pygmies of the Ituri Forest, the requirement for hierarchy is reduced to a minimum, and very stable egalitarian societies are possible. However, as the size of the groupings increases so must the requirement for hierarchy.

Each society has a whole set of beliefs regarding the supernatural forces that can be exploited to enable individuals, associations and society as a whole to achieve their ends. Many ceremonies and rituals are performed to this end, all of which have the additional effect of tightening social bonds, and hence of further increasing social organization. At the same time, every society has a set of taboos, basically to prevent supernatural forces from being mobilized to hinder the achievement of the society's goal-structure (7).

There is every reason that the goal that self-regulating societies set themselves is one whose achievement permits the satisfaction of the environment's many competing requirements and is not purely arbitrary, as in the case of our society. This may be illustrated by the way in which the size of the simple society is determined. Thus if the Eskimos live in small family units during the summer months, it is because there is no need for a larger unit, indeed the Arctic areas they inhabit would not support one. If the Pygmies of the Congo live in small bands, it is because this is the ideal number of people for survival in tropical rain forests, possibly providing the minimum number of hunters required to trap an elephant. If the society is truly self-regulating, however, it should be capable of reducing or increasing complexity to permit adaptation to changing environmental conditions – so long as these occur within certain limits. Thus when faced with the Macedonian menace, it would have been adaptive for the Greek city-states to join together to form a league, i.e. to achieve a higher level of organization. This they never really succeeded in doing, though there were many attempts.

On the other hand, in the absence of environmental challenges requiring action on the part of a larger, more complex social unit, it would be adaptive for complexity to be reduced, for the society to break up, temporarily at least, into its constituent parts. Usually, however, institutional barriers prevent this from occurring. Central governments are jealous of the territories that they control and usually refuse to face reality when environmental conditions render superfluous and artificial the states that they control.

The important thing is that a self-regulating society must be goal-directed. It moves in a particular direction, and both the goal towards which it is moving, and that behaviour pattern that permits its achievement, are culturally determined.

THE INTEGRITY OF A CULTURAL PATTERN

For the society to keep moving in this direction, it means that all its members must be imbued with the cultural information that will enable them to fulfil their specific functions as specialized members of their social system. It also means that every cultural trait which we often tend to regard as being of little practical significance, and which our missionaries, educators, administrators, etc., are only too pleased to interfere with, has a specific function in the overall social behaviour pattern.

If one were acquainted with the culture of any stable society and were capable of working out the role played by each of the customs and institutions within this culture, i.e. by determining in what way they contributed towards the adaptive behaviour of the society to its particular environment, one could easily imagine what would be the consequences of their suppression by outside interference. Let us take the case of the marital customs of the Comorians, who inhabit a group of islands between Mozambique and Madagascar. The people of the Comores have a complex social organization, probably based on indigenous customs upon which were superimposed those of their Islamic conquerors. From the former they inherited a matrilineal and matrolocal tradition; from the latter a patrilineal and patrilocal one. Islamic marital law has also been adopted. As a result, there is polygamy and a high frequency of divorce. Indeed, so high is the latter that it is perfectly normal for a woman to have been married five to ten times. From the experience gained in our society, we would tend to associate such a consequent number of 'broken homes' with a very high rate of juvenile delinquency, schizophrenia and suicide, i.e. the symptoms of social disorder. However, things do not work out that way.

In Mayotte, one of the islands making up the Comores Archipelago, there have been only two deaths by violence in the last fifty years, and neither was premeditated murder. Crime in general is minimal, as are mental diseases, delinquency, suicide and the other symptoms of social disorder.

The society is thus culturally adapted to marital instability, which ours is not. The reasons are two-fold. First, by virtue of the institution of matrilinearity and matrilocality, a child is partly the responsibility of the mother's clan. Many of the functions of fatherhood are in fact fulfilled by the mother's elder brother, and inheritance, for instance, is primarily through him rather than through the father. Secondly, by custom, the step-father automatically assumes many of the responsibilities of fatherhood *vis-à-vis* the children that his new wife has had with previous husbands. The stepfather, or 'baba combo', is, in particular, responsible for the payment of the very large expenses involved in the circumcision ceremony of his stepsons. Also, the father's role is reduced by the fact that the children are brought up in the mother's home. As the father probably has several other wives, he will in any case be physically present in one particular house on only one or two days a week. For all these reasons, divorce does not have the same unsettling effect in the Comores as it does in our society. Now, supposing a missionary or admin-istrator suddenly decided that matrilinearity and matrilocality were vestiges of barbarity not to be found in modern civilized societies, and that they must therefore be abolished; unless he abolished at the same time many of the other customs making up this complex culture, the results would be disastrous. Schizo-phrenia, delinquency, and the other symptoms of social dis-order would result, as they do in our society with the break-up of the nuclear family.

What is particularly striking about the self-regulating society is the absence of these forms of deviation. Crime is, in fact, an extremely rare occurrence, in spite of the fact that there are no policemen, law courts, tribunals, etc. Indeed in such a society, there is no need for external controls of this sort. Systemic controls, i.e. those controls applied by the society as a whole through the medium of public opinion, are sufficient to prevent any deviation from the accepted norm (4). As Linton writes:

The Eskimoes say that if a man is a thief no one will do anything about it, but the people will laugh when his name is mentioned. This does not sound like a severe penalty but it suffices to make theft almost unknown. Ridicule will bring almost any individual to terms, while the most stubborn rebel will bow before ostracism or the threat of expulsion from his group.

Besides, in a stable society all the citizens will have the good of the society at heart. They will feel part of it and will all equally oppose the behaviour on the part of any of its members that is contrary to the established customs and that might compromise the interests of the society as a whole. Solon was once asked which was the best policed city (8). 'The city', he replied, 'where all the citizens, whether they have suffered injury or not, equally pursue and punish injustice.' The same spirit as Solon expressed is apparent in Pericles' celebrated speech over the bodies of the first victims of the Peloponnesian war: '. . . we are prevented from doing wrong by respect for authority and for the laws, having an especial regard for those which are ordained for the protection of the injured, as well as for those unwritten laws which bring upon the transgressor of them the reprobation of the general sentiment.'

POLITICAL INSTITUTIONS

We regard government, parliaments, and vast bureaucracies as essential to all societies. However, it is probably that most of the societies ever developed by man dispensed with such external controls. Lowie writes (9):

. . . it should be noted that the legislative function in most primitive communities seems strangely curtailed when compared with that exercised in the more complex civilisations. All the exigencies of normal social intercourse are covered by customary law, and the business of such governmental machinery as exists is rather to exact obedience to traditional usage than to create new precedents.

Indeed, in such societies, nothing can be found to correspond to our notion of government. There are no kings, presidents, or even chiefs, no courts of law, prisons or police force. The closest approximation to a political institution is the council of elders that occasionally gathers to discuss important issues. It is for this reason that the Australian aboriginal tribe has often been referred to as a 'gerontocracy', or as a government by the old men – a title that can aptly be applied to most simple, ordered societies.

The absence of formal institutions, rather than giving rise to the permissiveness that we would expect, is in fact associated with discipline and the strictest possible adherence to the tribal code of ethics. Behaviour which in a disordered society could be exacted only at the cost of brutal coercion is with them ensured by the force of public opinion, the sanction of the elders, and the fear of the ancestral spirits.

In more advanced societies, we find the same principle obtaining in a less extreme form. Thus, in ordered societies where public opinion plays an important role, the need for strong government and, in particular, dictators is correspondingly reduced. Conversely, in those disordered societies where public opinion plays but a small part, we find that the absence of the most authoritarian government, linked to an all-pervasive and coercive bureaucracy, inevitably leads to lawlessness and mob rule.

By causing the disintegration of a society, by overloading the social system with too many people, or by increasing mobility so as to prevent their proper socialization, one is reducing the power of public opinion, and thereby the society's capacity for self-regulation. We are introducing asystemic controls in ever greater quantities: more politicians, more bureaucrats, more laws, more tribunals; so one is rendering the systemic controls ever more redundant, and further reducing a society's capacity for self-regulation.

A society used to being run in this manner ceases to be capable of running itself. In many South American republics, the deposition of one dictator will lead merely to the installation of another. Real democracy is not possible since the essential social structure for rendering it possible does not obtain. A mass democracy is, in fact, a contradiction in terms, and as our society becomes ever more massive and ever less organized, i.e. as entropy and randomness increase, so must there be a proportionate increase in the precarious asystemic controls required to maintain a semblance of social order, and a similar reduction in its stability and hence in its capacity to survive.

We can thereby formulate the essential principle that the higher the entropy or randomness of a social system the greater must be the need for asystemic controls, of which the most extreme kind is a dictator.

Such controls are as unsatisfactory in social systems as they are in ecosystems, and for the same reasons. A dictator gears his society to the achievement of what is usually an arbitrary goal regardless of environmental requirements. This can therefore only increase the society's instability. As we have seen, it creates a need for further dictatorship by destroying natural systemic controls which it renders redundant.

It is also highly vulnerable. One can exterminate a large proportion of the population of a self-regulating society without affecting its organization or its capacity to govern itself, whereas in a dictatorship it suffices to kill the dictator for the society to be plunged into chaos and civil war, a point that is particularly well illustrated by the experience of the later Roman Empire. Dictatorship, in other words, involves a drastic simplification of a society's control mechanisms, and must determine a corresponding reduction in its stability.

Our industrial society is further affecting human society by absorbing non-industrial cultures. Whereas there were previously innumerable cultures geared to totally different ends, so, as they fall within the orbit of industrial 'civilization', they come more and more to resemble each other.

It is significant that in New Guinea, that last great reservoir of primitive cultural wisdom, there are still 700 different cultural patterns, each with its own distinct language. In such circumstances mistakes committed by one such social group are likely to have the minimal effect on the others, and the probability that at least one cultural pattern provides the solution to a new environmental problem is maximized.

The absorption of these diverse highly differentiated societies into a common mass-society geared to the achievement of short-term material ends is a loss to humanity that cannot be overemphasized.

It must seriously reduce the complexity and hence the stability of human social organization on this planet. It must also lead to the irreparable loss of a vast store of cultural information which is as important to man's survival as is the store of genetic plant variety so seriously compromised by modern agricultural techniques.

SOCIAL DISRUPTION AND ITS EFFECTS

There is increasing evidence that deprivation of a satisfactory family environment will affect children profoundly and colour every aspect of their later life (10). Such children are often referred to as emotionally disturbed. However bright they may be, they will tend to find it very difficult to fit into their social environment, the reason being that the early and most important stages of socialization were badly impaired. The earlier family deprivation occurred, the more will this be the case, for, as D. O. Hebb (11), shows, the effect of early experience on adult behaviour is universally correlated with age.

Sadly, it is rarely possible for socially deprived and emotionally disturbed children to be satisfactorily socialized. No amount of school education can do much for them. They are characterized by their inability to accept any social constraints. They are unable to concentrate on their work and are interested only in things which are of apparent immediate advantage to them. Regardless of their intelligence level, they are thus extremely difficult to educate. They are particularly concerned with the present, and the short term, and are predisposed to all pathological forms of behaviour such as delinquency, drug addiction, alcoholism and schizophrenia. What is worse, when they grow up they are unlikely to be capable of fulfilling their normal family functions; their children, consequently also deprived of a normal family environment, will in turn tend to be emotionally unstable.

John Bowlby went so far as to compare a delinquent with a typhoid carrier (12). He is as much a carrier of disease as the latter – of a disease of the personality which will affect his family and his community for generations, until his descendants are eliminated by natural selection.

Socially deprived, emotionally disturbed youths are a feature of disintegrating societies. In the black ghettoes of New York and other large American cities, they are the rule rather than the exception. There is increasing reason to suppose that the low standard of achievement and the high rate of crime, and various forms of retreatism that characterize such societies, are mainly attributable to family deprivation.

If a child is seriously affected by being deprived of a satisfactory family environment, an adult is also adversely affected by being deprived of a satisfactory communal environment.

As we have seen, in a stable society a cultural pattern provides an individual with a complete goal-structure and an environment within which these goals can be satisfied. In a stable society the principal goal appears to be the acquisition of prestige, to be looked up to by one's family and community. In our industrial society, prestige is achieved in a variety of ways, including the right education, entering a socially acceptable profession and, perhaps most important of all, making money.

Those who have not been subjected to the normal socialization process, and in particular members of different minority ethnic groups, may for various reasons find these avenues of success barred to them. In such conditions they have no alternative but to develop a substitute set of goals. Cloward and Ohlin (13) interpret the development of a criminal sub-culture in the slums of a big city in these terms. It provides people with a new set of goals which they can achieve. Once crime becomes big business, and requires the same sort of qualities as permit success in the mainstream culture, than a further substitute outlet is required.

It is in these terms that Cloward and Ohlin interpret the 'violent gang' sub-culture, which also has its own ethic and goal-structure, so different from the mainstream culture. However, those who have not succeeded in shedding the latter's values find themselves incapable of participating in it. They are forced to indulge in one or other form of retreatism – to isolate themselves psychologically from an environment which not only fails to provide them with an essential set of goals but also denies them. Merton (14) describes a retreatist in the following way:

Defeatism, quietism and resignation are manifested in escape mechanisms which ultimately lead him to 'escape' from the requirements of the society. It is thus an expedient which arises from the continued failure to near the goal by legitimate measures and from an inability to use the illegitimate route because of internalised prohibitions, this process occurring while the supreme value of the success-goal has not yet been renounced. The conflict is resolved by abandoning both precipitating elements, the goals and the means. The escape is complete, the conflict is eliminated and the individual is associalized.

In a disintegrating society one would tend to find sub-cultures developing along all these different lines in varying degrees, i.e. there will be an increase in delinquency, violence and all the various forms of retreatism, such as drugs, drink, strange religious cults, etc., and mental disease. Such a society will be characterized by a general feeling of aimlessness, a frantic, almost pathetic search for originality, over-preoccupation with anything capable of providing short-term entertainment, and beneath it all a feeling of hopelessness of the futility of all effort.

CRIME

In the United States, according to Mr John Mitchell, Attorney-General, crime in cities of more than 250,000 inhabitants is two and a half times that of the suburbs, which in turn is twice that of rural areas (see Table 9). Crime, needless to say, is on the increase. In the United States it has doubled in the last ten years. In 1969 there were 2,471 crimes per 100,000 inhabitants. There were 655,000 violent crimes and 4,334,000 crimes against property, 14,590 murders, 36,470 rapes and 306,420 aggravated assaults. This is an increase of 12 per cent over the previous year. In the United Kingdom, crime is increasing at a similar rate. In 1970, according to a *Newsight* investigation, there were $1\frac{1}{2}$ million indictable crimes, 300,000 in London alone, an increase of about 10 per cent over 1969.

Crimes of violence and burglary and battery in particular are increasing at the fastest rate, at more than 15 per cent per annum. There are at present 66 crimes of violence per 100,000 people in the United Kingdom, as opposed to 324 per 100,000 in the United States. At the present doubling rate of five years it will take approximately 12 years to achieve the US rate of

324 per 100,000, which is so bad that life in cities has become intolerable and economic activity seriously menaced.

Professor Michael Banton of the Department of Sociology, Bristol University, told the British Association for the Advancement of Science that 'increased disorder is part of the price we pay for the adaptation of our social arrangements to an economic

Table 9   Increase in US crime rates with increase in size: urban crime rates per 100,000 population, 1957.

	Size of Cities		
	Over 250,000	50,000 to 100,000	Under 10,000
Criminal homicide			
Murder, non-negligent manslaughter	5·5	4·2	2·7
Manslaughter by negligence	4·4	3·7	1·3
Rape	23·7	9·3	7·0
Robbery	108·0	36·9	16·4
Aggravated assault	130·8	78·5	34·0
At this point the scale changes			
Burglary, breaking or entering	574·9	474·6	313·3
Larceny-theft	1,256·0	1,442·4	992·1
Auto theft	337·0	226·9	112·9

Reproduced by courtesy of the Pemberton Publishing Co., from *Population Versus Liberty* by Jack Parsons.

system which brings us such great material benefits'. Crime is part of the price of affluence or, more precisely, of the social disintegration that affluence gives rise to.

Perhaps the most damning indictment of our industrial society is the behaviour of people when the elaborate mechanisms of the law are for some technical reason put temporarily out of action. In Montreal, during a 24-hour police strike shops were pillaged, women raped and houses burgled. In London, during a power strike theft increased to such an extent in shops and department stores that many had to close until the light came on again.

Nothing better illustrates what can happen when the self-regulating mechanisms which normally ensure the orderly behaviour of the members of a stable society break down and are replaced by a precarious set of external controls.

ILLEGITIMACY

As the family unit breaks down, it is not surprising to find that illegitimacy, another symptom of social disintegration, increases. Nor is it surprising to find that it is closely linked with other symptoms of social disintegration. According to W. R. Lyster, an Australian statistician, 'Crime and illegitimacy rates are simultaneous in their incidence. The illegitimacy rate in England and Wales per 100 of all births has increased since 1955 from 4·7 to 7·8; crime has increased from about 45 per 10,000 to 120 per 10,000; thus, both have more than doubled.'

Illegitimacy is costing our government £52 million per year. In industrial slums and other societies that have reached the more advanced stages of disintegration it is not unusual to find that up to 70 per cent of children are illegitimate.

W. A. W. Freeman, President of the Children's Officers Association, has recently reported a startling increase in the number of women who are simply abandoning their children, something which would not occur in a stable society.

ALCOHOLISM

For each specific cultural pattern there must exist an optimum degree of alcohol consumption. It is likely that increases over and above this level will be in direct proportion to the development of disorder within the society itself. The number of offences of drunkenness proved in England and Wales for the year 1967 is greater than the number of offences proved in previous years. The increase occurred as expected in the large cities, the City of London having 476·43 offences for each 10,000 of its population. The Home Office, with characteristic ignorance of basic sociological matters, writes, 'No reason for the increase can be adduced. There was no significant change in the liquor licensing laws.'

According to the National Council on Alcoholism, alcoholism is costing the country about £250 million a year, mainly because of absenteeism from work. About seven workers out of every thousand have drinking problems, and there are about 400,000 alcoholics in the country, a figure which is increasing annually.

MENTAL HEALTH

Social disintegration is a major cause of mental disease. When an individual deprived of his essential social and physical environment is incapable of building a substitute one, or fails to isolate himself from the one he can no longer tolerate, by means of drugs or alcohol, his behaviour pattern, no longer adaptive to an environment for which it was not designed, tends to break down. One remaining position of defence is to build up his own personal world of fantasy which contains just those environmental constituents of which he has been deprived and which he most requires.

There is considerable evidence to show that members of a society undergoing acculturation, whose culture is breaking down under the influence of an alien one, are particularly prone to mental disease.

As national boundaries break down, small communities are swallowed up by vast urban conglomerations, mobility is increased and people move about the place in search of better pay, so cultural patterns break down.

In the United Kingdom, mental disease is increasing at a phenomenal rate. According to Ministry of Health statistics 169,160 people were admitted to hospitals in England and Wales in 1967 suffering from mental illness, two and a half times as many as in 1951.

There were 600,000 mentally disordered people in England and Wales in 1967, 186,901 of them occupying hospital beds or 46·6 per cent of all hospital beds. Thirty-two million working days every year are lost because of mental illness, representing a cost to the nation of £100 million, and local authorities spent £20,250,000 in mental health, more than six times what was spent in 1957.

SUICIDE

Durkheim regarded suicide as the ultimate manifestation of anomie. He found that the suicide rate was particularly low in poor rural communities where social structures were intact, and high in disintegrating affluent societies, especially among the working classes and even more so among immigrants, in this case Italians to the cities of Lorraine. He goes so far as to say that 'suicide varies in inverse proportion to the degree of integration of the social groups to which the individual belongs'.

In Britain the suicide rate has fallen over the last six years by about 200 a year. Nevertheless, according to the Samaritans, a lay organization that helps depressed and potentially suicidal people, the number of potential suicides has more than doubled in the last two years. In 1967 their seven London area branches dealt with 5,999 new cases. In 1969 the same branches dealt with a further 11,641 cases. The Reverend Basil Higginson, an official of this organization, estimates that cases would go on rising at this rate.

CONCLUSION

There is every reason to believe that the social ills at present afflicting our society – increasing crime, delinquency, vandalism, alcoholism as well as drug addiction – are closely related and are the symptoms of the breakdown of our cultural pattern, which in turn is an aspect of the disintegration of our society. These tendencies can only be accentuated by further demographic and economic growth. It is chimeric to suppose that any of these tendencies can be checked by the application of external controls or by treating them in isolation, i.e. apart from the social disease of which they are but the symptoms.

It is the cause itself, unchecked economic and demographic growth, that must be treated. Until such time as the most radical measures are undertaken for this purpose, these tendencies will be further accentuated – until their cost becomes so high that further growth ceases to be viable.

Appendix C:
Population and food supply

It is a common assumption that throughout the entire history of mankind human populations have expanded whenever conditions permitted. Thus it is argued that during the 100,000 generations in which our forebears lived by food-collecting the difficulties of keeping body and soul together were so great that populations were limited largely by crude food availability. Then with the adoption of agriculture, some 200 to 300 generations ago, the new-found sources of food permitted populations to expand until generally speaking they were held down only by disease. Finally, modern public-health methods, principally greatly improved sanitation and vector control procedures, permitted the phenomenal increases which collectively are known today as the population explosion.

Yet there is now good evidence that many of the human societies living before the agricultural revolution (and a few after) were stable societies in the strict sense of the phrase: i.e. they were regulated not by starvation, disease or war, but by cultural controls which only now we are beginning to understand. The basis of such controls is the individual sense of responsibility. Among the bushmen, for instance, the woman will avoid having a second child while she is suckling her baby; in the inhospitable environment in which she lives, it would be extremely difficult to look after both at once. In France, up to the Second World War, the population was relatively stable, principally because people were imbued with a strong sense of responsibility. A man would not get married until he was in a position to look after his wife, nor would he have children until he thought he was capable of bringing them up properly. When the motivation is there, the means tend to be found for satisfying it. The provision of advanced methods of contraception is of little use to a people who do not have the cultural controls necessary to maintain a stable population. In the Philippines, for instance, women tend to want more than six children. What use can it be merely to provide them with the pill?

Since in a nutshell the problem of populations and food supply is how to live within one's ecological means without being forced to do so by naked hunger, it is worth bearing in mind that man (like many other animals) is potentially capable of so doing. In the meantime, we are faced with the task of reducing birth-rate to compensate for the fall in death-rate, because, daunting though it undoubtedly is, the alternative of satisfactorily feeding an expanding population is still more so. Nonetheless, so far attempts to reduce birth rates have been largely ineffective on a global scale and even if they were to be successful it is unlikely that they could produce significant reductions in population growth rates within the time scale required to avoid major food shortages.

It is argued that the raising of living standards will, of itself, limit population growth by offering economic incentives that are reduced if the ratio of wage-earners to dependants within the family group is weighted too heavily in favour of dependants. Evidence of this is contradictory, although it is true that families have become smaller in Europe as levels of material prosperity have risen. However, it is not possible that this situation can be repeated throughout the world as a whole. The planet lacks the resources to permit the industrialization that would be required and even if these were to be found the levels of pollution generated by such a level of industrial activity would be greater than could be absorbed by the ecosphere. Even then, although the rate of increase would be reduced, populations would continue to grow.

The population of Britain is growing at 0·5 per cent per year, which gives it a doubling time of 138 years. While this is much lower than the world average (1·9 per cent each year), each individual within an industrial society consumes far more resources and contributes far more to environmental pollution than an individual in an agrarian society. Professor Wayne Davis (1) considers that an American has twenty-five times the impact on the environment as an Indian so that, worked out in terms of 'Indian equivalents', the population of the United States is equivalent to that of 5,000 million Indians. Thus the problem of population is more acute in developed than in developing countries.

We must consider whether it is possible for the planet to provide food in sufficient quantities to sustain the populations that are forecast.

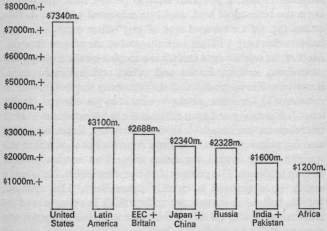

Yearly increase in
population expressed
in terms of GNP

Figure 4. This chart is one way, a very approximate way, of looking at the need for population control from an ecological standpoint. The longer the bar, the greater the need. The bar-lengths are arrived at by combining annual increase in population with the relevant *per capita* GNP. (In the United States, for example, population is increasing at about 2 million a year. *Per capita* GNP is approximately $3,670. The total population/GNP growth factor is therefore $7,340 million. Comparative figures for India are: yearly population increase, 13·4 million; *per capita* GNP, $90; total growth factor, $1,206 million. Include Pakistan, and this final figure swells to $1,600 million.) The rationale for this is the close tie-up between wealth and pollution.

Food production may be increased either by extending the area under cultivation or by intensifying production on existing farmlands, or by both. Current FAO programmes concentrate on intensifying production on existing farms.

The extension of agriculture into marginal lands is expensive in terms of investment and produces only limited returns. It is more rational to direct such capital as is available into the improvement of existing farming.

Indeed, the amount of marginal land available for agriculture is severely limited, and it has been estimated that if the required increases in food production were to be met from this source alone the reserves of land would be exhausted within a decade or less (2). Of a total land area of 32·5 billion acres, estimates indicate that only 3 billion are cultivated at the present time (3). Most of the world's land surface is occupied by the ice-caps and permafrost, deserts, forests and urban and industrial areas. Sometimes it is suggested that the remaining tropical forests, in Amazonia in particular, might be cleared to provide agricultural land. It is unlikely that such schemes could be successful, even if the resources were available to carry them out. Experience with clearing primeval forest in Central America has shown that the removal of the climax vegetation triggers an erosion process leading to desert. The process is all but irreversible, for organic matter once exposed is quickly mineralized. The unstable lateritic soils of Amazonia are 70 feet thick but they would be likely to erode very quickly if they were unprotected against the equatorial climate, while this itself would certainly be affected by the removal of such a large area of forest. When Khrushchev cleared the forests in Kazakhstan for agriculture he left a dust bowl of some 30 million acres, an area equivalent to the entire agricultural land area of the British Isles.

The US President's Science Advisory Committee estimated in 1967 that the total arable and potential arable land in the world amounted to 8 billion acres. While some expansion is possible, it is unlikely that the resources of capital and materials can be made available to produce more than minor increases in food production from these sources. There is little marginal land remaining for development in the Soviet Union, China, Asia or

Europe, and extension of farmlands in the more arid regions of the Middle East and North Africa would require new sources of fresh water for irrigation that are not available at present and will not be within the immediate future. It is possible that the United States might increase its area under cultivation from 300 to 350 million acres (3).

In fact, it is likely that existing agricultural land will be reduced as demands for urban and industrial development, with all that that implies in terms of roads, airports, railways, etc., are met. Between 1882 and 1952 the total land area of the world occupied by permanent buildings increased from 0·87 billion to 1·6 billion hectares (2). This will be much higher if, by the year 2000, 81 per cent of the population of the developed countries and 43 per cent of the population of developing countries will be living in urban areas (Table 10).

Beyond a certain point, which varies with climate and soil type, the intensification of farming causes soil deterioration and eventually erosion. This is already a problem in many of the developed countries, where very intensive farming systems have been imposed. The extension of monocultural arable farming, the heavy use of artificial fertilizers, the use of heavy machinery and, in other areas, overstocking with farm animals, all contribute to deterioration in soil structures, leading to a loss in the efficiency of drainage systems and in the effectiveness with which soluble fertilizers can be used. In this situation irrigation can lead to problems of waterlogging and/or salinity, while the over-consumption of groundwater for irrigation purposes can lead to a lowering of water tables that may compromise the future of farming. In large parts of Texas, for example, the present long drought is exacerbated by low water tables, and it is possible that farming in Texas may have to be abandoned altogether.

The deterioration of soil structure has been observed in Britain, where stable soils and a temperate climate provide near-ideal farming conditions. In more severe climates and on poorer soils erosion is likely to appear more quickly and once it begins it could become an accelerating process. As the poorer lands fail the pressure on the better lands will increase, so tending to

Table 10 Changes in land utilization 1882–1952.
(*In billion hectares*).

	1882	Percentage	1952	Percentage	Change 1888–1952	Percentage
Forest	5·2	45·4	3·3	29·6	−1·9	−36·8
Desert and wasteland	1·1	9·4	2·6	23·3	+1·5	+140·6
Built-on land	0·87	7·7	1·6	14·6	+0·73	+85·8
Pastures	1·5	13·4	2·2	19·5	+0·7	+41·9
Tilled land	0·86	7·6	1·1	9·2	+0·24	+24·5
	9·53	83·5	10·8	96·2	+1·27	+12·9
Area not especially utilized	1·81	16·5	0·27	3·8	−1·54	−79·9
Total	11·34	100	11·07	100	−0·27	−2·4

Reproduced from G. Borgstrom, *Too Many*.
Source: R. R. Doane, 1957. *World Balance Sheet* (Harper, New York).

encourage still further intensification which will damage soils more rapidly than might be anticipated (see Table 11).

Table 11   Quality classification of tilled land.

	1882 Percentage	1952 Percentage
Good	85·0	41·2
Half of original humus lost	9·9	38·5
Marginal soils	5·1	20·3

Reproduced from G. Borgstrom, *Too Many*.
Source: R. R. Doane, loc. cit.

Erosion of farmlands in some areas is associated with the pread of deserts. In 1882 the world had a total 1·1 billion hectares of desert and wasteland. In 1952 the area had increased to 2·6 billion hectares (2) (see Table 11).

Table 12   World average rates of increase for the period 1951–
           66 for selected aspects of human activity related to
           food production.

	Percentage*
Food	34
Tractors	63
Phosphates	75
Nitrates	146
Pesticides	300

Reproduced from SCEP.
Source: Digested from United Nations, *Statistical Yearbook*, 1967.
*Rates in constant dollars.

Given that demand for land must increase with population growth, and that populations are increasing exponentially, and assuming that the *per capita* requirement of land is 0·4 hectares for agricultural purposes and 0·08 hectares for non-agricultural purposes (a low estimate), Meadows (4) has shown that by the year 2000 the land available is likely to have decreased by 250 million hectares, while the demand will have increased by about 2·4 billion hectares, and that somewhere between 1980 and 1990 the demand for land will exceed the supply. Furthermore, if yields per acre were to double, the effect would be to add no more than 30 years to the world's food supply. Similarly a quadrupling of yield, which no serious person would consider possible, would add only 60 years. The net demand for food, then, will double every 30 years, and it can be satisfied only by doubling yield every 30 years.

Britain has one of the most intensive farming systems in the world. In the twenty-five years since the end of the Second World War very large sums of money have been invested in technological developments aimed at increasing output and reducing the requirement for labour. Nevertheless, when the effect of inflation on farm prices is taken into account, the productivity of British agriculture has increased by only 35 per cent, and there is good reason to suppose that in most major products yields have now levelled off and in some they are declining. Short of major technological breakthroughs in plant

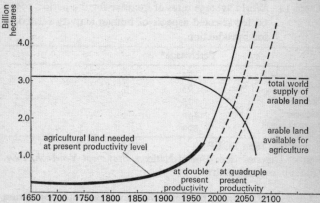

Figure 5. Arable land.

Total world supply of arable land is about 3·2 billion hectares. About 0·4 hectares per person of arable land are needed at present productivity. The curve of land needed thus reflects the population growth curve. The light line after 1960 shows the projected need for land, assuming that world population continues to grow at its present rate. Arable land available decreases because arable land is removed for urban-industrial use as population grows. The dotted curves show land needed if present productivity is doubled or quadrupled.

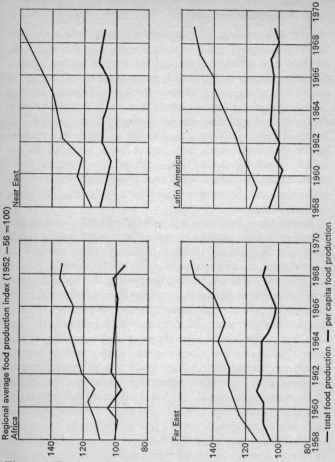

Figure 6. Food production.

Regional average food production index (1952–6 = 100)
Total food production in the non-industrialized regions of the world
has risen at about the same rate as the population. Thus food production *per capita* has remained nearly constant, at a low level.

Source: UN Food and Agriculture Organization, *The State of Food
and Agriculture 1970* (Rome, UN Food and Agriculture Organization, 1970).

genetics and, possibly, the introduction of entirely new concepts in farming, none of which is in sight at present, it is extremely unlikely that agricultural production in Britain can achieve further significant increases. It is not possible for agriculture in developing countries to receive the heavy investments that British agriculture has received, and so it is unlikely that increases in production can be achieved to match those in Britain. Even if they were, they would be insufficient, even to sustain the present inadequate dietary levels. Although the so-called 'Green Revolution' has produced important improvements locally, overall the world food situation shows no sign of improving, and there seems little chance of the FAO's targets for 1985 being met.

In past years local emergencies have been alleviated by the provision of food, principally grains, from world stocks, which have been held mainly in North America. These stocks have been allowed to run down and so even this 'cushion' is lost.

There are definite biological reasons for the limits on food production. Plants depend on a complex mixture of inputs, many of which are beyond man's control. Even of the principle requirements – sunlight, water and nutrients – it is only nutrient that man has succeeded in manufacturing and supplying to his crops. Fertilizer use is subject to diminishing returns beyond certain levels of application, and these may be much lower in the field than controlled experiment under near-laboratory conditions would suggest. Thus an 11 per cent increase in the agricultural production in the United States between 1949 and 1968 was achieved with a 648 per cent increase in the use of nitrogen fertilizer, while Britain's 35 per cent increase required an 800 per cent increase in nitrogen fertilizer consumption. The consumption of pesticides to control the effects of the ecological imbalances created by the farming system has increased also. Between 1950 and 1967 US pesticide consumption increased by 267 per cent and achieved a 5 per cent increase in total crop yields (6).

The use of agrochemicals on a large scale makes a serious contribution to the pollution of the global environment. They are biologically potent, which is why they are used, and when introduced at random into the environment they interfere with living processes. Many pesticides affect the central nervous system of man, they may interfere with hormone secretions, and some are known to be carcinogenic or teratogenic. Under certain circumstances some fertilizers can be harmful to health, and by forming random associations with amines present in the environment, nitrites can become nitrosamines, which are carcinogenic. There is no way of knowing the extent to which the environmental carcinogen and mutagen loads have been increased, because it is impossible to monitor all the possible interactions between pollutant and pollutant and between pollutants and substances present naturally. Pesticides that are persistent accumulate along food chains, so depressing predator populations and, in the long run, tending to encourage increases, rather than decreases, in pest populations. Organochlorine insecticides are particularly harmful to fish. Excess fertilizers enter water systems where they contribute to eutrophication problems. There must be an upper limit to the tolerance of the ecosphere to pollution from agriculture.

The effectiveness of pesticides is further reduced because insects, weeds and micro-organisms acquire resistance to them. Such resistance is based on hereditary characteristics in certain individuals within populations. It is transmitted genetically, and so repeated application leads to the build-up of a resistant pest population by a process similar in all ways to natural selection. Throughout the world there are now some 250 species of insect pest that are immune to most insecticides (7).

In common with most organic chemicals, pesticides are derived from petroleum, and their continued production is related to the availability of petroleum or of an alternative source of raw materials, although any alternative is likely to be more expensive. All agrochemicals consume power and water in their production, and the availability of cheap sources of power and plentiful supplies of water is likely to limit any increase in production.

The intensification of agriculture in many areas of the third world would require much-improved systems of transport to convey fertilizer, pesticides, machinery and seeds in and food out. It is doubtful whether the capital is available to develop such transport systems or the fuel to power them.

It is unrealistic to suppose that there will be increases in agricultural production adequate to meet forecast demands for food, and the notion that technological inputs can be made available that would guarantee a doubling of production by 1980 and a further doubling by 2100 is no more than fantasy. Such a thesis can be advanced only by 'experts' who fail to take into account basic ecological, physical and biological principles, or who are not in possession of all the relevant information.

The intensification of agriculture cannot prevent famines within the next fifteen to twenty years, probably affecting parts of Asia, Africa, the Near East and Latin America. Indeed, by causing further disruption to terrestrial and marine ecosystems it must reduce the capacity of the planet to support life.

Attempts to increase the world's protein availability from fisheries show no sign of solving the problem. The seas are experiencing serious pollution which may be undermining the phytoplankton that form the base of the marine biotic pyramid, and they may be overfished. In 1969, for the first time in a quarter of a century, total fisheries production did not increase, owing to poor catches, and this in spite of heavy capital investment by the developed countries. Fishery vessels are operating in deeper and more remote waters and owing to the high level of investment in ships and processing plant the developed countries, which are also the major fishing nations, are irrevocably committed to increasing yields by a large factor within a very short space of time. In their efforts to do so there is little reason to suppose that they will not so deplete fish stocks that within a decade or so the contribution of fisheries to world food supplies will reduce rather than increase. If there is a temporary increase, little of this will benefit the developing countries, which by and large cannot afford to participate in such a heavily capitalized operation. At present less than 20 per cent of the world's total catch of sea and fresh water fish is consumed within the third world.

Appendix D:
Non-renewable resources

INTRODUCTION

For the purposes of this discussion, non-renewable resources are
divided into two types: metals and fuels.

METALS

The 16 major metals we are concerned with are:

Silver (Ag)
Aluminium (bauxite) (Al)
Gold (Au)
Cobalt (Co)
Chromium (Cr)
Copper (Cu)
Iron (Fe)
Mercury (Hg)
Manganese (Mn)
Molybdenum (Mo)
Nickel (Ni)
Lead (Pb)
Platinum (Pt)
Tin (Sn)
Tungsten (W)
Zinc (Zn)

As can be seen from Figure 2 on p. 24, at present rates of
consumption all known reserves of these metals will be exhausted
within 100 years, with the exception of six (aluminium, cobalt,
chromium, iron, magnesium and nickel). However, if these rates
of consumption continue to increase exponentially at the rate

they have done since 1960, then all known reserves will be exhausted within 50 years with the exception of only two (chromium and iron) – and they will last for only another 40 years!

Of course this is by no means the whole picture: there will be new discoveries and improvements in mining technology, and we can turn to recycling, synthetics, and substitutes. It should be obvious, however, that recycling, although a necessary and valuable expedient in a stable economy, cannot supply a rising demand (it is not a source of metals, merely a means of conserving them); while synthetics and substitutes cannot be imagined into production, but must be made from the raw materials available to us, those most suitable being themselves in short supply. Petroleum, for example, from which many valuable synthetic polymers are derived, will run out within the lifetime of those born today and will probably be increasingly scarce – and correspondingly expensive – from about the year 2000. Improvements in mining technology will be necessary in any case if we are to make use of the lower grades of ore that will be the only ones available to us as reserves are depleted. However, exponential increases in consumption will inevitably lead to a situation in which grades decline much faster than technology is improved and costs will therefore soar. Similarly, as William W. Behrens (1) has shown, the dynamic of exponential growth will considerably reduce the lifetime of new discoveries. For example, even if reserves of iron (which has a relatively long lifetime) are doubled, they will stave off exhaustion for only another 20 years. Thus, given present rates of usage and the projected growth of those rates, most raw materials will be prohibitively expensive within about 100 years. Political difficulties will arise well before then – as indeed they are beginning to do in the case of oil.

As Preston Cloud (2) has pointed out, the extra iron, lead, zinc and so on necessary to raise the level of consumption of the 3,400 million non-Americans to that of their fellows in the United States is from 100 to 200 times present annual production – and, although this would be exceptionally difficult to achieve, it is paltry compared with the problem of providing an equivalent standard of consumption for the doubling of world population projected for 40 years' time. And yet we in the industrial countries expect our consumption of metals to go on rising and at the same time lure the non-industrial countries with promises that they too can have 'wealth' like ours!

Only those acrobats of the imagination who argue that, come what may, technology will find a way believe that problems such as these can be solved in any way save by a diminution of consumption. In particular, they are confident that the abundance of cheap energy they assure themselves will be available in the near future will enable us to extract the metals present in ordinary rock and in sea water. Yet energy is already very cheap (comprising only 4·6 per cent of the world's total industrial production by value) (3), while the real limit on such enterprises is likely to be not energy but the fragility of ecosystems. For example, the ratio of unusable waste to useful metal in granite is at least 2,000:1, so that the mining of economic quantities of metals from rock or sea water will very quickly burden us with impossible quantities of waste.

ENERGY

The bulk of our energy requirements today is met by fossil fuels, which like metals are in short supply. At present rates of consumption, known reserves of natural gas will be exhausted within 35 years, and of petroleum within 70 years. If these rates continue to grow exponentially, as they have done since 1960, then natural gas will be exhausted within 14 years, and petroleum within 20. Coal is likely to last much longer (about 300 years), but the fossil fuels in general are required for so many purposes other than fuel – pesticides, fertilizers, plastics, and so on – that it would be foolish to come to depend on it for energy (4).

Recognition of this has led to the present emphasis on nuclear fission as a source of energy. However, the only naturally occurring, spontaneously fissionable source of energy is uranium-235, and this is likely to be in extremely short supply by the end of the century (5). Accordingly, the future of nuclear power rests with the development of complete breeding systems. Breeder reactors use excess neutrons from the fission of uranium-235 to convert non-fissionable uranium-238 and thorium-232 into fissionable plutonium-239 and uranium-233 respectively. Their successful development will mean that man's energy needs will probably be met for the next 1,000 years or so, during which time it is hoped that deuterium-deuterium fusion can be developed – which will provide us with virtually unlimited energy.

Because the successful development of breeder reactors in time to take over from fossil fuels is possible, it may be that fuel availability will not be a limiting factor on growth. This means nothing, however, since shortages of other resources and pollution by radioactive by-products and waste heat will quickly prevent the continued expansion of energy consumption. Since radioactive pollutants have been dealt with in Appendix A on ecosystems, we will here consider only waste heat.

Every use of energy always produces waste heat. Power stations 'solve' the problem of heat production by using large amounts of either cooling water or, to a lesser extent, air. The disadvantage of the former method is that if the heated water is returned to source it damages the aquatic ecosystem, and if it is evaporated into the atmosphere the source is considerably depleted. The disadvantage of the latter method is that because air temperatures are higher than those of water the thermo-dynamic efficiency of the power station is much reduced.

Efficiency is a great problem. In the US, electricity provides 10 per cent of the power actually used by the consumer, but accounts for 26 per cent of gross energy consumption. Earl Cook (6) has calculated that, at present rates, by the year 2000 electricity will provide 25 per cent of 'consumer-power' and account for between 43 and 53 per cent of gross energy consumption. At that point, half the energy produced will be in the form of useful work and half in the form of waste heat from power stations.

Even if we ignored the waste heat from power stations, that produced by the actual consumption of electricity will quickly call a halt to growth. For example, in the US in 1970 heat from that source amounted to an average of 0·017 watts per square foot, and Claude Summers (7) has calculated that if consumption continues to double at the present rate, within only 99 years, after 10 more doublings, the average will be 17 watts per square foot – compared with the average of 18 or 19 watts the US receives from the sun! Clearly, well before this point energy consumption will be limited by the heat-tolerance of the eco-sphere.

The Movement for Survival (MS)

I AIM

We need a Movement for Survival, whose aim would be to influence governments, and in particular that of Britain, into taking those measures most likely to lead to the stabilization and hence the survival of our society.

2 STRUCTURE

We envisage it as a coalition of organizations concerned with environmental issues, each of which would remain autonomous but which saw the best way of achieving its aims was within the general framework of the *Blueprint for Survival*.

These organizations have already expressed general support for the *Blueprint*:

The Conservation Society
Friends of the Earth
The Henry Doubleday Research Association
The Soil Association
Survival International

Two representatives of each member organization would join the Action Committee of the MS, which would elect a chairman and secretary to run the day-to-day business of the Movement.

3 INDIVIDUAL MEMBERSHIP

Members of constituent organizations would automatically become members of the MS. Individuals who belonged to none of these bodies could join the MS only through one of these organizations.

Regular news of MS activities would be published in *The Ecologist*, a subscription to which would be available to MS members at the reduced price of £3 p.a. (25 per cent reduction). Representatives of the constituent organizations could become members of *The Ecologist*'s editorial board.

4 FURTHER INFORMATION

Organizations wishing to join the MS and all others seeking further information should write to the Acting Secretary, The Movement for Survival, c/o *The Ecologist*, Kew Green, Richmond, Surrey.

References

INTRODUCTION: THE NEED FOR CHANGE

1. Jay Forrester, *World Dynamics*. Wright Allen Press, Cambridge, Mass, 1970.
2. FAO, *Provisional Indicative World Plan for Agriculture*. Rome, 1969.
3. Stated by the Ministry of Agriculture to the Select Committee on Science and Technology, *Population of the United Kingdom*. London, HMSO, 1971.
4. Agricultural Advisory Council, *Modern Farming and the Soil*. London, HMSO, 1971.
5. Ministry of Agriculture (Statistics Division), *Output and Utilisation of Farm Produce in the United Kingdom, 1968-9*. London, HMSO, 1970.
6. The American Metal Market Co., *Metal Statistics*, 1970; P. T. Flawn, *Mineral Resources*, 1966; Dennis Meadows *et al.*, *The Limits to Growth*, 1972; United Nations, *The World Market for Iron Ore*, 1968; United Nations, *Statistical Yearbook, 1969*, 1970; US Bureau of Mines, *Minerals Yearbook*, 1968; US Bureau of Mines, *Commodity Data Summary*, 1971; *Yearbook of American Bureau of Metal Statistics*, 1970.
7. Dennis Meadows *et al.*, *The Limits to Growth*. London, Earth Island, 1972.

TOWARDS THE STABLE SOCIETY

1. WHO, *WHO Chronicle*—special issue on DDT. Geneva, 1971.
2. Barry Commoner, 'The Causes of Pollution', *Environment*. March/April 1971.
3. Sir Otto Frankel *et al.*, 'The Green Revolution: genetic backlash', *The Ecologist*. October 1970.
4. Robert Allen and Edward Goldsmith, 'The Need for Wilderness', *The Ecologist*. June 1971.

5. Kenneth Boulding, 'Environment and Economics', *Environment*, ed. William W. Murdoch. Stamford, Conn., Sinauer Associates, 1971.
6. National Academy of Science, Natural Resources Council, *Recommended Dietary Allowances*. Washington, 1953.
7. Stephen Boyden, 'Environmental Change: Perspectives and Responsibilities', *Journal of the Soil Association*. October 1971.

THE GOAL

1. John Stuart Mill, *Principles of Political Economy*. Vol. II. London, John W. Parker, 1857.
2. René Dubos, 'Can Man Adapt to Megalopolis', *The Ecologist*. October 1970.
3. 'Energy Slaves', *The Ecologist*. January 1970.
4. Edward Mishan, 'The Economics of Hope', *The Ecologist*. January 1971.
5. Bishop of Kingston, *Doom or Deliverance ?* Rutherford lecture, 1971.

APPENDIX A: ECOSYSTEMS AND THEIR DISRUPTION

1. C. H. Waddington, *The Strategy of the Genes*. London, George Allen & Unwin, 1957.
2. See Bryn Bridges, 'Environmental Genetic Hazards', *The Ecologist*. June 1971.
3. *SCEP*, *Man's Impact on the Global Environment*. M.I.T. Press 1971.

APPENDIX B: SOCIAL SYSTEMS AND THEIR DISRUPTION

1. Paul R. Ehrlich, 'Hobson's Choice', in *The Optimum Population for Britain*, ed. R. Taylor. Academic Press, 1970.
2. N. Tinbergen, *The Study of Instinct*. Oxford, Clarendon Press, 1951.
3. G. P. Murdock, *Social Structure*. New York, The Free Press, 1965.
4. Ralph Linton, *The Study of Man*. London, Peter Owen, 1965.
5. Jose Ortega y Gasset, *España Invertebrada*.
6. Daisy Bates, *The Passing of the Aborigines*. London, John Murray, 1966.

7. See Abram Kardiner, *The Psychological Frontiers of Society*. New York, Columbia University Press, 1945.
8. Thucydides, ii, 37. Quoted by W. Warde Fowler, *The City State of the Greeks and Romans*. London, Macmillan, 1952.
9. Robert Lowie, *Primitive Society*. London, Routledge, 1952.
10. Marshall B. Clinard, *Anomie and Deviant Behaviour*. New York, The Free Press, 1964. This book contains a very useful bibliography accompanied by abstracts of the books dealing with the whole subject treated in this article.
11. D. O. Hebb, *The Organisation of Behaviour*. New York, John Wiley, 1961.
12. John Bowley, *Child Care and the Growth of Love*. Harmondsworth, Penguin Books, 1965.
13. Richard E. Cloward and Lloyd E. Ohlin, *Delinquency and Opportunity*. New York, Collier-Macmillan, 1966.
14. Robert K. Merton, *Social Theory and Social Structure*. New York, The Free Press, 1967.

APPENDIX C: POPULATION AND FOOD SUPPLY

1. Wayne Davis, 'Four Billion Americans', *The Ecologist*. July 1970.
2. G. Borgstrom, *Too Many*. New York, Collier-Macmillan, 1970.
3. Lester Brown and Gail Fintserbusch, 'Man, Food and Environment', *Environment*, ed. William Murdoch. Sindner Associates, 1971.
4. Dennis Meadows, *The Limits to Growth*. London Earth Island, 1972.
5. Michael Allaby, 'The World Food Problem', *Can Britain Survive?*, ed. E. Goldsmith. Tom Stacey Ltd, 1971.
6. Barry Commoner, 'The Environmental Cost of Economic Growth', paper presented at Resources for the Future Forum. Washington, 1971.
7. *Environment* staff report, 'Diminishing Returns on Pesticides', *Can Britain Survive?*

APPENDIX D: NON-RENEWABLE RESOURCES

1. William W. Behrens, 'The Dynamics of Natural Resource Utilisation', *Proceedings of the 1971 Computer Simulation Conference*. Boston, Mass.
2. Preston Cloud, 'Mineral Resources in Fact and Fantasy', in *Environment*, ed. William W. Murdoch. Sinauer Associates, 1971.

3. United Nations, *Statistical Yearbook*, 1969.
4. *World Petroleum Report*. New York, Mona Palmer Co., 1968 .
5. M. King Hubbert, 'The Energy Resources of the Earth', *Scientific American*. September 1971.
6. Earl Cook, 'The Flow of Energy in an Industrial Society', *Scientific American*. September 1971.
7. Claude M. Summers, 'The Conversion of Energy', *Scientific American*. September 1971.

More about Penguins and Pelicans

Penguinews, which appears every month, contains details of all the new books issued by Penguins as they are published. From time to time it is supplemented by *Penguins in Print*, which is a complete list of all available books published by Penguins. (There are well over three thousand of these.)

A specimen copy of *Penguinews* will be sent to you free on request, and you can become a subscriber for the price of the postage. For a year's issues (including the complete lists) please send 30p if you live in the United Kingdom, or 60p if you live elsewhere. Just write to Dept EP, Penguin Books Ltd, Harmondsworth, Middlesex, enclosing a cheque or postal order, and your name will be added to the mailing list.

Note: *Penguinews* and *Penguins in Print* are not available in the U.S.A. or Canada

A Pelican Book

Only One Earth

The Care and Maintenance of a Small Planet

Barbara Ward and René Dubos

Man has been washed up on an island, like Robinson Crusoe. How is he to survive?

Only One Earth sets the key for the United Nations Conference on the Human Environment to be held at Stockholm in June 1972. The report, which is unofficial, has been read and in parts revised by more than 150 expert consultants from many countries and many fields. Logical in order and cogent in expression, this extraordinary document is a Domesday Book of the kingdom of man.

Here is earth's swelling population; here are its measurable resources; here are the knowledge, energy, industry and commerce that form man's potential; and here, too, in the squalid details of technology's impact on soil, sea and air, is man's record in fouling his own nest.

All this amounts to one thing for the industrialized nations and another for those with plans to develop. And yet, as Barbara Ward and René Dubos so movingly argue, it is only one earth that man inhabits.

What must he do to be saved?

Not for sale in the U.S.A.